STEM EDUCATION WITH ROBOTICS

This book offers a synthesis of research, curriculum examples, pedagogy models, and classroom recommendations for the effective use of robotics in STEM teaching and learning. Authors Chauhan and Kapila demonstrate how the use of educational robotics can catalyze and enhance student learning and understanding within the STEM disciplines.

The book explores the implementation of design-based research (DBR); technological, pedagogical, and content knowledge (TPACK); and the 5E instructional model; among others. Chapters draw on a variety of pedagogical scaffolds to help teachers deploy educational robotics for classroom use, including research-driven case studies, strategies, and standards-aligned lesson plans from real-life settings.

This book will benefit STEM teachers, STEM teacher educators, and STEM education researchers.

Purvee Chauhan earned a Master's in Education degree from Harvard University, USA, with a focus on Technology and Innovation. She has 10 years of field-experience in the education sector, has held global impact positions in organizations such as New York University (NYU), Teach for India, and The Nalanda Project, and is currently an Instructional Designer at ASCD, one of the largest education non-profits in the United States.

Vikram Kapila is a professor of mechanical and aerospace engineering at the NYU Tandon School of Engineering, USA, where he directs a Mechatronics, Controls, and Robotics Lab. His current research is focused on the convergence of frontier technologies (e.g., robotics, artificial intelligence, augmented/virtual reality, and blockchain) with applications to human-robot interaction, digital health, and STEM education.

STEM EDUCATION WITH ROBOTICS

Lessons from Research and Practice

Purvee Chauhan and Vikram Kapila

NEW YORK AND LONDON

Designed cover image: Used assets from Freepik.com

First published 2023
by Routledge
605 Third Avenue, New York, NY 10158

and by Routledge
4 Park Square, Milton Park, Abingdon, Oxon, OX14 4RN

Routledge is an imprint of the Taylor & Francis Group, an informa business

Library of Congress Cataloging-in-Publication Data
Names: Chauhan, Purvee, author. | Kapila, Vikram, author.
Title: STEM education with robotics : lessons from research and practice / Purvee Chauhan and Vikram Kapila.
Description: New York, NY : Routledge, 2023. | Includes bibliographical references and index.
Identifiers: LCCN 2022056568 (print) | LCCN 2022056569 (ebook) | ISBN 9781032367569 (hardback) | ISBN 9781032367576 (paperback) | ISBN 9781003333616 (ebook)
Subjects: LCSH: Robotics in education. | Robotics in education—Research—Case studies. | Science—Study and teaching. | Technology—Study and teaching. | Educational technology—Research. | Educational technology—Research—Case studies.
Classification: LCC LB1028.76 .C43 2023 (print) | LCC LB1028.76 (ebook) | DDC 371.33—dc23/eng/20221220
LC record available at https://lccn.loc.gov/2022056568
LC ebook record available at https://lccn.loc.gov/2022056569

ISBN: 978-1-032-36756-9 (hbk)
ISBN: 978-1-032-36757-6 (pbk)
ISBN: 978-1-003-33361-6 (ebk)

DOI: 10.4324/b23177

Typeset in Bembo
by Apex CoVantage, LLC

To my mother who had bigger dreams for me than I could ever imagine and to my husband for his relentless support and encouragement throughout this journey.
– Purvee Chauhan

To my parents and brothers for their constant support in my academic journey and to my wife for her unconditional love and understanding.
– Vikram Kapila

CONTENTS

FIGURES

Part I

Chapter 1

Chapter 2

Chapter 3

Part II

Chapter 4

Chapter 5

Chapter 6

Part III

Chapter 8

Chapter 9

TABLES

Chapter 4

Chapter 6

Chapter 8

Chapter 9

Appendix A

Appendix B

FOREWORD

In the early 1920s, a young Czech playwright, Karel Čapek, was at work on a play. Up to that point, he had minor success as a journalist and a political critic, and this was his first foray into writing for the theater. It was a play based in the future, a piece of science-fiction long before science-fiction had become an established genre. In this somewhat dystopian play, Karel described a day at a factory populated by artificial people, created from synthetic organic matter, who took care of most of the work typically done by humans. Since these beings were crucial to the story he was trying to write, he needed to find an appropriate name to describe them. He struggled to devise a name for these workers, initially considering calling them *labori*, from the Latin *labor*, for work. Not satisfied with that neologism, he turned to his brother and frequent collaborator, Josef Čapek, for inspiration. Josef suggested the word *roboti* (which also meant labor or hard work in the Slavic languages). Karel liked the suggestion, thus, the word *roboti* ended up in the title of his play: RUR or *Rossumovi Univerzální Roboti*. The play premiered at the prestigious Prague National Theater on January 25, 1921, to positive reviews, and was soon translated into multiple languages and performed across the world. It premiered in New York City in October 1922, where none other than Spencer Tracy (the soon-to-be Hollywood superstar) played a word-less robot. Karel went on to write more plays, stories, and other politically engaged works in favor of free expression and in opposition to fascism and communism. He was nominated seven times for the Nobel Prize in Literature, though he never won the award. He died at home of pneumonia in 1938 just before he was about to be captured by the Gestapo for his anti-fascist views. His brother, the person who actually coined the word *roboti*, was not as lucky, and died in a concentration camp.

We must recognize that despite his literary fame when alive, Karel Čapek's work is barely remembered today. And the one thing he is remembered for is coming up with the word *robot*. Though there had been earlier descriptions of human-like automatons, and Čapek's creations are more akin to what we would now call androids, the word, robot, has stuck—particularly once it entered the English language in translation as *Rossum's Universal Robots*.

Robots soon become a mainstay of science fiction, whether in books or in cinema, for humanoid machines that can do various tasks for humans. But robots are no longer just fiction. Though they may not (for the most part) look humanoid, programmable mechanical devices that perform tasks without the aid of human interaction are omnipresent in our world today. From the Roomba vacuums zooming around our homes, to the machines that automatically harvest corn; from robotic hands that help surgeons conduct intricate surgeries, to the dancing robots created by Boston Dynamics—robots are already an important part of our present, and even more so, of our future.

The science and technology of robotics design, manufacturing, and application has grown by leaps and bounds over the past decades. This revolution has been fueled by developments both in software (particularly machine learning and artificial intelligence) as well as better understanding of the mechanics of movement and object manipulation. In that sense, robotics is almost a perfect encapsulation and convergence of ideas in the STEM (science, technology engineering, and mathematics) disciplines.

As we look to the future, it becomes increasingly important that we understand the role of the STEM disciplines in defining and determining our future. And robots, by extension, will be an important part of this. Thus, since education is inherently about preparing our next generations for the future, we must consider ways to help youth in our schools and universities become more knowledgeable about what these technologies mean for us.

That said, one of the challenges of such technologies is their opacity—by which we mean that they are often "black boxes," with their inner workings hidden from us. By contrast, the inner workings of most industrial age tools (such as mechanical machines and clocks) were visible to the naked eye. The opacity of newer digital technologies often makes them appear impervious to human intervention and meaning-making.

Therefore, it is unsurprising that most depictions of robots and their influence on our lives, even starting with Čapek, have been dystopian. Science fiction has often picked up on fears about the potential wrong directions or dystopias brought about by unknowingly diving into technological phenomena with our eyes wide shut. We rarely have any understanding of the inner workings of complex digital artifacts, or any sense of how things are constructed and come together. The result of this lack of understanding is often a sense that technological phenomena are often happening *to* us. There is a lack of awareness, from a critical perspective, about how such technological tools are engineered to shape our lives or behaviors, for both better and worse. People are often left feeling that they simply must trust in the tools to work properly and to shape society in our best interests, rather than feeling like they have an actual stake in the game. At both individual and societal levels, people need a better understanding of STEM, developed through the use, construction, and manipulation of hands-on technological knowledge—such as robotics—to make the opaque workings more transparent.

There are social, ethical, and pedagogical imperatives to equipping our next generation of citizens to be more informed and thoughtful about how these technologies work. To understand how these tools function, we must appreciate and recognize that they are human-made creations that can be re-made, re-imagined, and re-designed to better serve broader humanistic goals. By developing STEM through robotics, students have the opportunity to explore the making and working of tools that are shaping our world, and which will certainly shape our future. Whether or not they go into the field of robotics, learners need opportunities to both build STEM knowledge through engaging modalities like robotics, and develop that sense of critical awareness of the inner workings of technologies.

A strong societal drive for STEM advancement and learning goes back to the start of the 'space race' that developed between nations back in the 1950s and 1960s. But since that time, the need for STEM development has grown and proliferated internationally. Nations often judge themselves in education by comparing STEM scores on tests like PISA. There is also much discussion and concern over the forecasted lack of STEM workers to meet the demand of the future. However, many of the discussions around STEM would do better to acknowledge the importance of engaging students in applications that are driven not only by workforce needs—but by STEM learning that actively connects to their desire to construct, to inquire, and to delve into exciting and constructive modes of thinking, making, and learning.

That is where this book's work on robotics learning through STEM can be so powerful. It delves more deeply into work that is both practical and scholarly, driven by theories such as technological, pedagogical, and content knowledge (TPACK), yet also grounded in classroom practice—both showing what creative and powerful robotics STEM learning can look like, and rigorously investigating its outcomes. Too often when we see instances of robotics learning in classrooms, it is either in one-off examples of studies published in a single article or chapter, or in specific practical examples of an interesting lesson. This book offers a more comprehensive exploration of the topic in a more in-depth, long-term, and thorough approach. The authors take us through their multi-year journey from its inception through the development over time. They ground their investigations in foundational theoretical frameworks, such as design-based research, TPACK, and the 5E instructional model. This both situates their work within educational research and extends it as well through the design of examples, lessons, and cases to present a picture of this long-term STEM learning robotics project. The concrete, detailed learning elements in many chapters have practical resonance for educators as well as the investigation of research outcomes. This book, in that sense, occupies a unique place in its presentation of STEM robotics learning. It offers much to both educational practitioners and researchers alike to help in developing teachers and learners who can engage and inquire within STEM in hands-on, understanding-driven ways.

This work is essential, if for no other reason than to prevent the dystopian vision of Karel Čapek (and many others) from becoming reality. It is a deeply humanistic effort and the authors of this book should be applauded for taking on this challenge through a multi-year, multi-faceted, and pedagogically impactful research program.

Punya Mishra and Danah Henriksen
Arizona State University

PREFACE

Even as today's students live in and benefit from a world enriched by scientific discoveries and technological advances, they lack a foundational understanding of science, technology, engineering, and mathematics (STEM) disciplines. For the United States (US) to maintain its competitive advantage in the innovation-driven global economy, leaders of education, government, and business have come to recognize the importance of fostering and sustaining early interest in STEM disciplines and the critical need for 21st-century skill development among K-12 students. Yet, the contemporary curriculum and instruction at the K-12 education level often lack a meaningful adoption of technology in the classroom or continue to focus on technocentric integration. Building engaging and active learning experiences using emerging technologies requires interventions at the curriculum, pedagogy, and teacher development levels. In recent years, educational robotics has gained popular attention from educators as it allows students to construct their own learning and understand abstract concepts through real-life problem-solving.

This book is a culmination of over two decades of education, research, training, mentoring, and outreach work to catalyze the incorporation of contemporary, modern technologies in K-12 educational environments. This work has entailed: creating STEM curricula that is intentionally and systematically infused with technologies, providing professional development opportunities to teachers, and enriching STEM learning experiences of students. Since 2002, under a series of federal-, state-, local-, and philanthropy-funded efforts, these educational activities have engaged over 500 teachers to integrate engineering concepts in the science and math classrooms and science labs of dozens of urban public schools in the northeastern US, impacting over 16,000 students. This STEM education research, conducted as a collaborative partnership involving engineering and education faculty, postgraduate and graduate researchers, and K-12 educators, has (1) created, implemented, and examined over 100 standards-aligned robotics-based science and math lessons and (2) developed, practiced, and examined research-guided pedagogical approaches for science and math learning using robotics. The resulting STEM programs take advantage of robotics as a tool to transition students' extra-curricular robotics experiences to an in-class setting for STEM learning. Moreover, through a series of research studies, this effort has investigated: *Whether the motivational power and new affordances of robotics can be effectively harnessed to positively influence the learning of science and math in the K-12 environment?*

This book, *STEM Education with Robotics: Lessons from Research and Practice*, seeks to synthesize and present research outcomes, curriculum examples, pedagogy models, and classroom recommendations from a multi-year teacher professional development effort that focused on middle school STEM education through robotics. This book is conceived as a vehicle for sharing newly gained knowledge, effective strategies for professional development, engaging instructional models, novel illustrations of STEM curriculum, and outcomes of research with teachers, researchers, teacher educators, and other stakeholders. Specifically, it caters to the lack of intentionally organized learning content at the intersection of robotics, STEM, professional development, curriculum, and education research. The book seeks to serve as a much-needed bridge between research and practice by providing the models, strategies, and lessons learned that have all been co-created by the researchers in partnership with teachers in real-world classroom settings.

Through a purposefully considered collection of examples and case studies, the book explores the implementation of several theoretical constructs, such as design-based research (DBR); technological, pedagogical, and content knowledge (TPACK); features of effective professional development; Bloom's taxonomy; and 5E instructional model, that enable effective use of robotics in STEM teaching and learning. The chapters in the book include a variety of scaffolding to prepare readers in effectively deploying robotics for classroom use, without compromising with the rigor of the content. Moreover, each case study includes a set of recommendations that can be adopted by the readers seeking to undertake similar explorations. Finally, each chapter enumerates a set of key takeaways to summarize its highlights. The book consists of the following three parts.

Part I: Introduction

Significant research was conducted during the planning, implementation, and assessment of the multi-year effort that became the basis of this book. Drawing from this research, this first part conveys the significance of the intersection of STEM education with robotics as well as the fundamentals of educational robotics and its myriad affordances, all to prepare educators for generating transformational STEM learning with robotics. Moreover, it reviews prior literature to highlight the foundational theories and instructional approaches relevant for robotics-infused learning. Several examples on the use of robotics in various roles and settings are also included. Finally, this part highlights the theoretical constructs, project design, and activities of the multi-year collaborative research effort for developing robotics-based STEM curriculum and a teacher professional development model.

Part II: Theory, design, and implementation

This second part focuses on describing the various theoretical underpinnings of the research project for curriculum design and teacher professional development. Specifically, it describes the frameworks of DBR and TPACK as well as the features of effective teacher professional development, all of which constitute the foundational elements of the collaborative research effort. For each of these concepts, this part provides the background material, their applications in varied settings with relevant examples, their purposeful integration in the design and implementation efforts, and the resulting outcomes. This coverage will profit teacher educators and education researchers seeking to design their own research-guided educational interventions.

Part III: Instructional perspectives and lesson designs

The third part of the book elucidates the alignment between the theoretical elements from Part II with the intended curriculum and pedagogical elements. Specifically, it first outlines the prerequisites and practices for effective integration of robotics in classroom teaching and learning and student perceptions regarding the same. Next, it highlights the applications of Bloom's taxonomy and the 5E instructional framework for creating robotics-infused experiential learning. These exemplars will benefit teacher educators and classroom teachers with various instructional and curriculum ideas pertaining to educational robotics and empower them to independently plan for their classrooms.

The nine chapters of this book are arranged in a logical manner for ease of comprehension and adoption of its content and strategies, making it a potentially valuable resource for teachers, researchers, and teacher educators alike. The book provides many illustrative applications of educational robotics and offers examples of various theoretical constructs that can lead to engaging, effective, and sustainable implementation of emerging technologies in the classroom. The book will appeal to educators and researchers seeking to learn about: effective integration of robotics in middle school STEM curriculum and instruction, teacher professional development, and application of research methods such as DBR and TPACK in STEM educational interventions. The book also elucidates recommendations that can help teachers in determining prerequisites, practices, and perceptions about instructional perspectives related to effective incorporation of robotics in STEM learning as well as aligning robotics-based lessons with prevailing science and math standards. Adapting the learnings from this book will enhance teacher effectiveness and capacity in teaching abstract STEM concepts in middle schools—through curriculum, aligned instruction, and hands-on learning activities; impact student interest in and motivation for learning STEM; and help make robotics central to and sustainable in middle school science and math classrooms.

ACKNOWLEDGMENT

This book is the result of more than two decades of grant support under the Discovery Research PreK-12 (DRK-12), Graduate STEM Fellows in K-12 Education (GK-12), Innovative Technology Experiences for Students and Teachers (ITEST), and Research Experiences for Teachers (RET) Site programs, all funded by the National Science Foundation (NSF), and the Central Brooklyn STEM Initiative, funded by six philanthropic foundations. Additional support for curriculum design, teacher professional development, and student learning enrichment from several state and local agencies is also acknowledged. These efforts have been collectively supported by more than 100 engineering students. The principal source of support for a bulk of the work reported in this book is based on an NSF-supported DRK-12 project under the award DRL: 1417769. The contents, findings, and recommendations delineated in this work are the sole responsibility of the authors and not reflective of the views of any sponsors.

The authors acknowledge the contributions of 43 teachers from 19 middle schools and their over 2,000 students who participated in the DRK-12 research study over a span of six years. This book would not have been possible without the utmost dedication and support of these teachers who co-created the educational robotics curriculum and professional development model as well as enacted the robotics-based STEM learning experiences for their students. Seven graduate researchers, six postdoctoral research associates, four faculty colleagues, and two staff researchers contributed to the work on curriculum, professional development, and research studies cited throughout the book. The authors specifically acknowledge the following research collaborators for reviewing various chapters of the book and providing constructive feedback for improvement: Drs. Ryan Cain, Sonia Mary Chacko, Shramana Ghosh, Saiprasanth Krishnamoorthy, and

Pooneh Sabouri. The authors also thank Dhruv Avdhesh for preparing some illustrations as well as Sidhaant Gupta and Hassam Khan Wazir for creating a few images of robotics artifacts for this book. Finally, the authors are grateful to five anonymous reviewers who provided constructive feedback on the draft book proposal including several sample chapters.

PART I
Introduction

The first part of the book seeks to involve its readers through three chapters to gain a strong foundation in the fundamentals of educational robotics, understand its significance, appreciate its connections with science, technology, engineering, and mathematics (STEM) education, and be exposed to its myriad applications. The first chapter, *Transformational Learning with Educational Robotics*, provides a detailed description of the intersection of robotics with integrated STEM and affordances of robotics in engendering active learning experiences. The second chapter, *Applications of Robots in Educational Settings*, builds on the fundamentals, through a detailed literature review, by introducing the foundational theories, instructional approaches, and roles and applications of robotics in varied educational settings. Finally, the third chapter, *Teaching STEM with Robotics: Synopsis of a Research-guided Program*, exposes the readers to the design and outcomes of a multi-year research effort that employed a wide array of theoretical constructs to implement a teacher professional development program and co-create novel robotics-based STEM curricula in partnership with teachers.

DOI: 10.4324/b23177-1

FIGURE PI.1 Learning with educational robotics.

1

TRANSFORMATIONAL LEARNING WITH EDUCATIONAL ROBOTICS

1. Introduction

Interdisciplinary learning originating from the four core subjects of science, technology, engineering, and mathematics, or STEM learning, is gaining increasing importance for success in the 21st-century global economy that is largely shaped by scientific discoveries and technological advances. STEM education helps learners in cultivating critical thinking and analytical skills through interdisciplinary explorations [1] and in developing a "STEM literate society" [2]. Moreover, such learning can prepare students to not only navigate the future of work but also fulfill the demanding needs of today's innovation economy. Emerging educational technologies have immense potential in helping STEM learners in advancing their scientific and mathematical aptitude [3]. Incorporation of technology in STEM education can enable authentic and active engagement, promote peer collaboration, and help students attend to learning tasks. However, not all technology-based education allows for active experiences, especially if the technological tools are employed to merely serve administrative purposes and fail to advance pedagogical needs [4].

Educational robotics, that is, meaningful integration of robots as learning tools, has attracted a substantial growth in use to facilitate interdisciplinary education in classrooms [5]. Hands-on, constructionist [6], and real-life activities of educational robotics have shown significant impact on student learning and skill development [7]. Educational robotics was introduced to increase student interest in technology and to involve them in topics that may be challenging to teach via traditional methods [8], thus facilitating a movement away from passive consumption or administrative use to a more active and instructional use of the tool. In addition to enhancing learning outcomes, educational robotics increases

DOI: 10.4324/b23177-2

student motivation [8, 9], fosters collaboration [8, 9], and promotes ownership and agency [10], among a myriad of other affordances. Moreover, teacher exposure and preparation to integrate robotics in curriculum can lead to interdisciplinary learning opportunities in STEM.

The chapter first discusses the importance and challenges of integrated STEM, that is, purposeful interactions between the four disciplines, in K-12 classrooms. Next, the chapter explores the role of educational robotics in designing STEM experiences wherein such an integration can allow students to engage in active learning with opportunities to collaborate, think critically, transform abstract concepts into tangible experiences, and gain a deeper conceptual understanding. Finally, the chapter introduces a case study that investigates the role of robots as artifacts in classrooms, revealing learning outcomes such as positive attitude, improved understanding, and enhanced engagement. In summary, the chapter aims to prepare educators in using educational robotics to produce transformational STEM learning experiences.

2. STEM and technology-enhanced learning environments

STEM, an acronym for science, technology, engineering, and mathematics, is an interdisciplinary intersection of these four closely related domains of study [11]. In STEM education, the disciplines of science, technology, engineering, and mathematics are purposefully integrated to create a "meta-discipline" that is undivided and "treated as one dynamic, fluid study" [1, 12].

Recent years have seen an exponential rise in STEM education in the United States (US). One key driver for this increase is the emphasis placed on STEM learning by the National Science Foundation (NSF), an independent US federal agency [13], through its initiatives that aim to provide students with critical thinking and problem-solving opportunities [14]. The result of improved STEM understanding during the formative K-12 stage can make students better equipped to attend college or train them for work if they opt to forgo post-secondary education [11, 14]. Under the purview of these initiatives, states and local education bodies have attempted to make STEM more inclusive by incorporating the "T" and "E" within the "S" and "M" offerings while also upgrading their technology education curricula [11]. Moreover, there has been an evolution of nationally developed disciplinary standards, and many states have sought to contextualize them for situations that are authentic and relevant to today's students [11]. NSF emphasizes the importance of studying STEM in the 21st-century as scientific discoveries and technological innovations become increasingly critical in global and knowledge-based economies [15]. They further assert that STEM capabilities of students need to be developed beyond conventional thresholds that existed in the past [15].

2.a. Integrated STEM: A framework

According to [16], STEM-integrated education entails students creating technological artifacts by purposefully learning and intentionally applying science and math through an engineering design process. Such an approach creates opportunities for students to see the relevance of and interconnections between science and math topics as well as how the concepts taught in school serve as foundations for technologies that they use in their lives. The STEM integration framework [16] advocates the inclusion of the following components in learning and exploration activities: engaging and motivating contexts; inclusion of engineering design; treating failure as a learning opportunity; intentionally employing science and math learning to address engineering problems; aligning learning tasks and activities to be student-centered; and promoting collaborative teamwork.

2.b. The importance of STEM integration

STEM is most meaningful and comprehensive when all the four subject areas are taught as an integrated whole [11]. Numerous studies have elucidated the wide array of knowledge, skills, and mindsets that students develop as they undertake STEM learning. Successful STEM interventions lead to enhanced student learning [2], instill 21st-century competencies [17], and focus on conceptual understanding [18] as opposed to rote memorization or superfluous use of step-by-step strategies. STEM learning is based on the pedagogical attributes of hands-on, real-life, and interdisciplinary investigations that help students in cultivating critical thinking, problem-solving, and analytical skills [1]. STEM education through constructionist making activities [19, 20] also allows learners to foster creativity and resilience [21] as they ideate multiple solutions, experiment, and iterate on feedback.

With an increased ability to transfer skills across the four disciplines, learners cultivate interest in STEM career pathways and develop STEM identities [17]. This interest in STEM careers is pivotal in a world that is increasingly becoming more technology focused. Moreover, to sustain and evolve a technology-driven economy, we need to develop a "STEM literate society" [2]. Bybee defines a STEM-literate individual as someone who possesses:

- "knowledge, attitudes, and skills to identify questions and problems in life situations, explain the natural and designed world, and draw evidence-based conclusions about STEM-related issues;
- understanding of the characteristic features of STEM disciplines as forms of human knowledge, inquiry, and design;
- awareness of how STEM disciplines shape our material, intellectual, and cultural environments; and

- willingness to engage in STEM-related issues and with the ideas of science, technology, engineering, and mathematics as a constructive, concerned, and reflective citizen" [2].

STEM learning has increasingly become a topic of dialogue, with urgent calls for adoption and implementation, not only to address educational demands but also to cater to the dynamic needs of the innovation economy and to fulfill community and business needs. When combined with the use of familiar and contemporary technologies [22, 23], and made personally and culturally relevant [24], STEM learning can generate enthusiasm and interest among students, in turn, building a more effective future workforce. Finally, these opportunities can act as a means of providing accessible and inclusive learning to students regardless of their demographics, socio-economic status, gender, or race.

2.c. Challenges in STEM integration

Even though there is an appreciation among educators to make STEM learning personally meaningful and relevant to students to prepare them for the future, the already overcrowded curriculum exacerbates attempts by teachers to integrate the four subjects as a purposeful whole. Currently, STEM is taught in K-12 classrooms in many varied ways, such as: the four subjects being taught independently with little or no integration and referred to as S-T-E-M; two out of four subjects being accorded greater emphasis, that is, more focus on science and mathematics than technology and engineering and hence, called SteM; or one subject being integrated in the other three subjects. For instance, engineering is taught through science, technology, and mathematics [11].

As evident, one common challenge of STEM education has been the seamless inclusion of "T" and "E," that is, "technology" and "engineering," with the other two subjects [14]. Since teachers have greater familiarity with science and mathematics, they feel a higher level of comfort in teaching these subjects, thus creating "educational silos" with little or no integration among the four study areas [14]. Understandably, without formal or sufficient training, teachers may feel apprehensive to teach engineering and/or technology. Moreover, they may not fully appreciate what engineers actually do and/or they may consider technology as referring merely to the field of computers. This limits their skills and abilities to prepare and deliver effective STEM lessons [14].

Another challenge in implementing STEM learning in schools is the absence of clarity among educators and administrators about the definitions and vision of STEM [1]. Furthermore, a deficit of STEM disciplinary knowledge and the inability to contextualize the use of contemporary technologies among teachers deter them from being effective in classrooms [25, 26]. Among students, this leads to apathy, inadequate preparation, poor engagement, and lack of skills in STEM disciplines [27] or failure to find value and meaning in what they are learning

[28]. Collectively, all these factors push students further away from developing a STEM identity and pursuing STEM pathways [17]. Moreover, the lack of effective collaboration among teachers limits the prospects for the success of STEM implementations [1]. Just as STEM learning spurs students to collaboratively brainstorm solutions, collaboration among teachers can help improve their lesson designs, promote peer learning, and help them practice their ideas [29]. All these findings illustrate the urgent need for raising awareness and preparing educators and schools about STEM-focused learning so that their students can thrive during their academic preparation and in the workforce.

2.d. Role of educational technology

One way to overcome the aforementioned challenges is by expanding learning opportunities through engaging, active, and immersive technology integration in STEM environments. Advancement of technology tools has impacted conventional classroom instruction through the adoption of various multimedia tools in lessons, incorporation of remote learning resources, ubiquity of richer research tools, accessibility of varied simulations, etc. [3]. Such an approach can lead to increased student motivation, learning engagement, personalization, and peer collaboration [3]. Technology can enable authentic learning by meaningfully simulating real-life scientific scenarios. Moreover, technology-supported learning can be designed such that users can focus their attention on one specific learning object at a time to reduce cognitive load and enable better understanding [3].

It has been well-established that technology-enabled educational strategies engender learning settings that center on students, create authentic and active learning experiences, and encourage students to chart their own paths [4]. Additionally, technology-infused learning can: support students in developing life skills such as self-regulation and problem-solving, improve learning outcomes, enhance their ability to visualize concepts, and lower their cognitive load [4]. Yet, in elementary and secondary school settings, resource and expertise constraints constitute a significant barrier for building and sustaining learning environments that effectively incorporate technology [4, 23].

A majority of teachers use technology for administrative and communication tasks as opposed to innovatively enhancing their teaching practices [4]. The mere presence or use of technology in a classroom does not guarantee its productive deployment to facilitate effective pedagogical techniques such as inquiry-based or active learning. According to [4], in one study over 50% of responding teachers indicated that they employed technology to ask students to complete homework using computers or use drill-and-practice software in classrooms. To create more student-centered hands-on learning experiences, teachers need to be trained and exposed to the varied approaches that draw upon the pedagogical potential of technology tools as opposed to focusing solely on the features of the tools [4]. Teachers will be more likely to effectively integrate a technology tool when they are able to visualize its appropriate use [30] and apply it to render concrete representations of abstract content [31]. It is essential to recognize teacher perceptions, attitudes, and specific requirements to support them in incorporating technological tools in learning [4, 30]. Following the substitution, augmentation, modification, and redefinition (SAMR) model, effective technology integrations are not limited to the use of technology as a direct substitution tool without any functional improvement; instead, they induce substantial redefinition of tasks, which were once unattainable, to engender authentic and engaging experiences [32].

Emerging technologies such as virtual reality (VR) and augmented reality (AR) can afford learners immersive STEM learning experiences [33, 34] that help in discerning complex disciplinary intersections in an engaging manner [3]. Modern technologies can also aid students in organizing, manipulating, and observing various scientific and mathematical phenomena in novel ways for enhanced learning [3]. For instance, by understanding a learner's pace of reading and comprehension, an artificial intelligence (AI) tool can individually guide her in perusing more rigorous texts. Tools such as sensors and electronics can help learners explore and experiment with real-life concepts and design solutions for their community. For example, learners may use pH and total dissolved solid sensors to assess the quality of water in their urban city apartment and then suggest various approaches for water remediation. Extending additional examples using such modern and accessible technologies can pave the way for a more purposeful, engaging, and meaningful learning and for the future of technological innovation.

3. Educational robotics

Technology in education can involve myriad tools such as computer-based learning, integration of immersive VR and AR technologies, game-based learning, maker tools such as 3D printers, robots, etc. The past several years have witnessed a significant growth in interest and efforts by educators in transforming interdisciplinary learning experiences with the use of robotics as a technology tool in classrooms [5]. Robotics is an interdisciplinary branch of engineering that integrates computer science, electrical engineering, electronics engineering, and mechanical engineering [35]. It entails designing, building, and using robots, all of which require applications of fundamental concepts of physics, mechanics, electricity, electronics, sensing, actuation, microcontrollers, computing, and engineering design, among many others. Robotics is increasingly being used in many real-world settings and it has found a plethora of applications in health care, manufacturing, security, entertainment, space exploration, and education.

The emerging field that uses robots as a learning tool is termed as *Educational Robotics* [8]. It "is an interdisciplinary learning environment based on the use of robots and electronic components as the common thread to enhance the development of skills and competencies" among learners [36]. Educational robotics, recognized for its value in developing cognitive and social skills of students, has the promise to serve as an effective contributor in bridging the STEM learning gap [5].

3.a. Affordances of using educational robotics

When educational robotics was first introduced, it had two primary learning goals: to enhance student interest in technology and to engage them in topics that are deemed challenging to treat using conventional methods [8]. Educational robotics is known to allow students to gain a deeper understanding about technology by meaningfully and interestingly applying their content knowledge and skills through hands-on activities [37]. Robotics can be used to teach multiple disciplines and it provides learners with opportunities to collaborate, think critically, and problem-solve [37]. Incorporation of real life-based active learning in educational robotics lessons can have a significant impact on student learning outcomes, including fostering 21st-century skills among them [7]. Application of robotics as a learning technology tool in classrooms is deemed to offer multiple affordances and, as delineated below, several researchers have sought to identify and unpack these in recent literature.

(i) Facilitates easy adoption and rapid exploration [38, 39]

Many students are familiar with robots from movies, science fiction, and other media [40]. Moreover, their perceptions of robots as toys can promote the adoption of educational robots for teaching and learning wherein robots can serve as a hook for student engagement in learning [41]. Educational robotics platforms are commonly conceived to facilitate ease of use for learners of varied age groups (see Figures 1.1 (a)–(d) illustrating preschoolers to high schoolers engaged in robotics activities). Children's experience with construction blocks used in childhood play imparts familiarity with building blocks of educational robots [42]. As students use educational robots in a classroom, their joyful memories of playing with construction toys may encourage them to engage in robotics-based learning activities [41]. The availability of a plethora of LEGO TECHNIC [43] and VEX [44] construction blocks enables learners to build and test various configurations of structures and mechanisms for their robotics creations. The electronic hardware elements of educational robots also support quick adoption and integration with their plug-and-play capabilities. Furthermore, educational robots equipped with a multiplicity of sensors allow learners to explore in varied ways the state of the robot and its environment. Finally, the now common visual programming environments of educational robots engage learners in rapid prototyping and testing of diverse solution strategies. The wealth of construction pieces, sensing and actuation devices, and coding blocks offered by educational robotics platforms afford opportunities to foster learners' creativity and enact diverse pedagogical strategies centered around learners. For example, thoughtful use of a robot's

(a) (b) (c) (d)

FIGURES 1.1 (a)–(d) Preschoolers to high schoolers engaged in interaction (a), exploration (b), experimentation (c), and programming (d) activities with robots.

sensors in experiential learning activities can draw upon students' understanding of an abstract concept or a textbook formula and highlight its real-world implication by linking it to a tangible measurement [45].

(ii) Supports scaffolding [46, 47]

The act of aiding learners to grow their cognitive and metacognitive awareness and behavioral skills is termed as scaffolding [41, 48]. The amount of support and its form, for example, social learning environment or concrete materials, can be a significant factor in the learning process of students [49, 50]. Beyond instructors or peers, with tools and artifacts to provide scaffolds [51], the materiality of student-created artifacts, for example, educational robotics, can foreground students' interactions with peers [47]. In fact, educational robotics offers myriad opportunities for supporting students to internalize the knowledge and abilities needed in completing given tasks [46]. For example, embedding exploratory robotics-related tasks in a learning sequence can test students' knowledge while clearly enunciated goals, which may be further

FIGURE 1.2 Middle schoolers using worksheets as a scaffold in a robotics-based lesson.

broken down by students, can prime their minds for robotics activities. As needed, students can be introduced to key vocabulary terms, scientific concepts, mathematical expressions, and engineering design approaches through hand-outs and multimedia tools and encouraged to use them throughout a robotics lesson (see Figure 1.2 illustrating middle schoolers using worksheets as part of a robotics-based lesson). To promote a deeper comprehension of concepts and methods, instruction and learning can be partitioned into smaller units and interspersed with hands-on explorations. Video and live demonstrations of robot activities can be used strategically to illustrate to students the expected operation and behavior of the robot. Students can be organized in small groups to support a social learning environment and allow for individual guidance. As students collaboratively brainstorm ideas, the instructor may probe individual groups to share their thinking and guide those who may be struggling [52]. Finally, as students progress through various stages of learning, the guidance and instruction may be tapered on the basis of the level of their observed understanding.

(iii) Imparts interdisciplinary learning opportunities [8, 9]

Robotics is an inherently interdisciplinary field that entails syntheses of knowledge, components, and subsystems from multiple domains in an integrative process to produce a whole that is larger than the sum of its parts. Analogously, educational robotics being essentially an interdisciplinary artifact [23] necessitates foundational knowledge of science and math to understand and enact engineering designs for producing technological solutions in response to real-world challenges. In this spirit, educational robotics possesses the promise to reveal to learners the interconnectedness of STEM disciplines and it has been championed to provide interdisciplinary learning of STEM disciplines [8]. Moreover, educational robotics can engender intrinsically authentic learning environments [10] where learners get opportunities to access and comprehend complex STEM concepts [31] and creatively express their understanding [8, 53]. For example, as students plan the movement of a robot traveling a specified distance, they have to devise a relationship between the distance traveled by the robot and the circumference of its wheels. Contrary to the general belief that robotics can be employed to improve learning only in STEM fields, it has been successfully incorporated for learning in myriad disciplines [8], for example, creative arts [53, 54], language learning [55, 56], social studies [57], etc. In fact, learners can participate in community-based projects to design a social robot to address stress in teens [58] or draw inspiration from a social robot that can accompany elderly population to reduce loneliness [59].

(iv) Increases student motivation [8, 9]

The theory of motivation deals with the factors that propel individuals to act and the two distinct factors are labeled intrinsic and extrinsic [60]. The use of educational robotics provides teachers with a means to motivate students intrinsically and extrinsically [61]. Specifically, the use of contemporary and novel educational robotics technology can induce self-interest among learners through freedom to engage in personally meaningful projects, create better products, or offer novel solutions, all of which can engender intrinsic motivation [7, 62, 63]. Alternatively, fascination with robots and group pride in popular robotics contests can transform an opportunity to learn with an educational robot into a reward to produce extrinsic motivation [62–64]. As a physical artifact, a robot promotes kinesthetic learning through hands-on manipulations that can contribute to student motivation [65]. Producing inviting and exciting learning environments [66], which infuse enthusiasm [67] and enjoyment [7] among learners through hands-on engagement, makes educational robotics a highly motivating learning tool [45, 62, 68]. Use of educational robots with students of both genders has been considered and deemed promising [56, 69, 70]. Moreover, educational robotics can increase STEM retention among female students and constitutes an equitable learning tool for students from diverse backgrounds and learning needs [8]. See Figures 1.3 (a)–(b) illustrating students participating in robotics activities with interest and enjoyment. As learners with varied interests in art, music, or storytelling, among others [5], partake in thought-provoking projects, they are presented with opportunities to

(a)

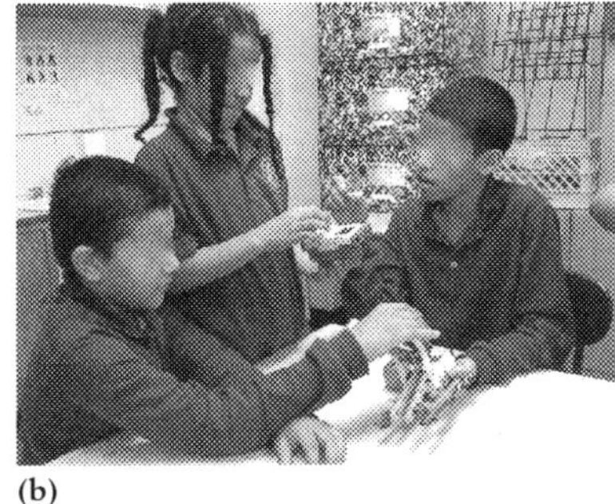

(b)

FIGURES 1.3 (a)–(b) Students partaking in robotics activities with interest and enjoyment.

create their own artifacts, build their knowledge, and take ownership of their learning [8, 66, 71, 72]. In such learning environments, students feel eager to do computations, learn new concepts, and share their work with peers [9, 67]. Moreover, robotics-enhanced education fosters student-initiated learning with personal relevance [5, 7, 9].

(v) Helps in understanding abstract concepts [8, 9]

When abstract STEM concepts are introduced with traditional instructional methods, in the absence of applications relevant to real life and lacking accessible representations, students find them challenging to comprehend [67]. Interactions with concrete artifacts through kinesthetic manipulation, visual observation, and auditory feedback [46] foster exploring and sense-making that can transform an abstract STEM concept into an accessible and tangible real-world understanding [7, 31, 69]. In this vein, by offering concrete, relatable, and often visual representations of knowledge [7, 31, 62, 69], educational robotics engenders an environment more conducive to learning [8]. See Figure 1.4, which illustrates the concepts of energy transfer and friction using a robotics-based mini golf lesson. Specifically, with the adoption of robotics as a learning tool, the instantaneous visual response helps students comprehend its input-output behavior. Any mismatch between the expected *versus* actual response of the robot prompts the students to explain the phenomena being probed and formulate a troubleshooting response [41]. Research has shown that students learned programming concepts such as conditionals, variables, and loops better with robots as they received instant, real-time feedback from the robot with the program execution [9]. Thus, as learners build, operate, and test their robotics creations, their progress and success in construction, integration, and programming choices [69] are revealed to them, encapsulating the status of their learning and reinforcing it [8].

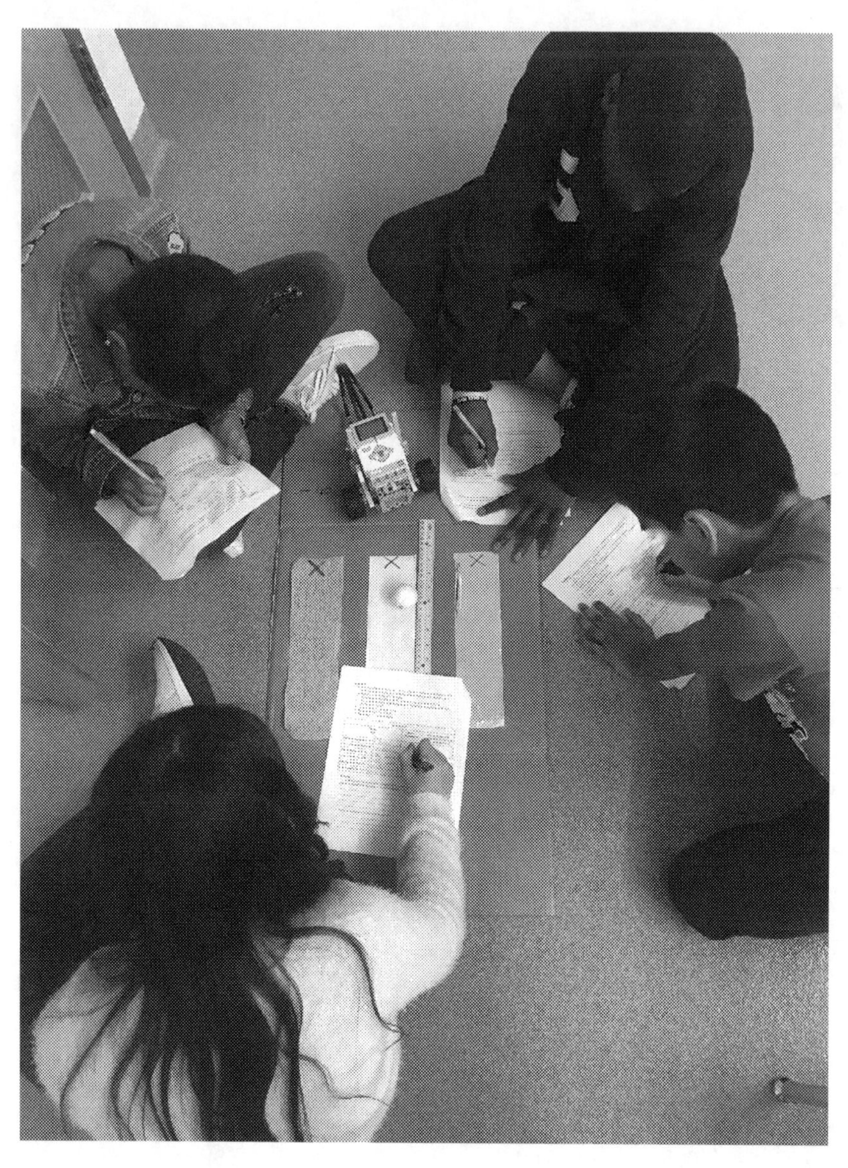

FIGURE 1.4 A robotics-based mini golf lesson to illustrate the concepts of energy transfer and friction.

(vi) Fosters collaboration [8, 9]

Use of collaboration as an instrument for knowledge building fosters active participation, close interaction, and productive dialogue among learners through which they can collectively identify areas for growth, share information, and solve problems [64], while also deriving the ancillary benefit of social-emotional development [73]. In a departure from didactic instruction, educational robotics activities are commonly structured as collaborative learning wherein students participate in team-based shared endeavors through learn-plan-design-test robotics explorations all while demonstrating impulse control and self-restraint [31] to produce a conducive environment for teamwork. In performing the varied robot building, programming, and integration tasks, students feel empowered to take ownership of the effort by taking turns and assuming the necessary roles such as team leader, assembly engineer, designer, programmer, inventory manager, etc. [67]. Through the course of the learning experience, students engage in collaborative decision-making, evaluating peer work, improving their design, iterating, etc. [8]. Moreover, working in groups has been shown to have a significant impact on the self-efficacy, collaboration, and communication skills of learners and, thus, is a great value add for educational robotics [9]. See Figures 1.5 (a)–(c) illustrating students engaged in robotics-based collaborative pursuits.

FIGURES 1.5 (a)–(c) Learners engage and collaborate on solving problems using educational robotics.

(vii) Imparts 21st-century skills and computational thinking [8, 9]

With the ever-changing work landscape, 21st-century skills are the attributes that an individual needs to embody and master to thrive and innovate in an increasingly complex work environment [74, 75]. Popularly, these skills include the 4C's of critical thinking, communication, collaboration, and creativity [7, 74–76]. Moreover, educational robotics literature [9, 77] suggests consideration of computational thinking, which entails "thinking or problem solving like a computer scientist" [78]. Educational robotics constitutes a formidable tool for cultivating both 21st-century skills and computational thinking due to its interdisciplinary, hands-on, and active learning nature [8]. Specifically, through a robotics-focused STEM education framework, students can (1) hone collaboration through learn-plan-design-test team activities; (2) practice communication skills through code comments and presentations; (3) engage in critical thinking via challenging, rewarding, and active project-based learning; and (4) foster creativity through design-build-program explorations via hands-on experiential learning. Programming being an integral step for implementing robotics-based learning activities, students are afforded multiple opportunities to formulate, program, test, and refine their code [8]. In coding robots for real-world challenges, students can practice *integrative learning*, for example, decomposing problems, recognizing patterns, analyzing solutions [77, 78], using math concepts [30, 66], and applying science and engineering practices [79], for example, formulating problems, modeling phenomena, designing solutions, etc. Such a strategy can catalyze learner creativity, reasoning abilities, systems thinking, and problem-solving skills, which form the foundation of computer science [78] and engineering, among other disciplines.

(viii) Enhances STEM learning [80]

By facilitating active experiences educational robotics offers a promise to transform the learning process, especially in STEM fields. Studies on educational robotics have shown substantial improvement in academic outcomes in STEM content areas that are intimately aligned with formal coursework [45, 80]. In addition to supporting the exploration of science and math principles [45, 81], robotics activities can help embed the engineering design process in STEM learning [82] and address the interconnectedness of STEM disciplines in the spirit of the Next Generation Science Standards (NGSS) [83]. The ability to conduct hands-on experimentation with educational robotics offers opportunities that can support students gain an understanding of abstract STEM concepts through concrete real-world applications [80]. Such an approach renders STEM education more accessible to students with diverse learning needs and generates interest among them. Moreover, the integration of educational robotics and STEM can draw on student interest and allow them freedom of expression as they address personally relevant challenges with social implications. Doing so makes STEM education more inclusive by engaging female students and students from underserved groups [54, 69, 84] and it has the potential to motivate learners in pursuing STEM coursework and careers in their future [80].

(ix) Promotes design of varied learning settings [85]

To serve the learning needs of today's diverse students, learning science research suggests creating and enacting four learning settings that center on the learner, knowledge, assessment, and community. Even as each of these learning settings has its own unique attributes, it is best considered as a component of one interconnected learning setting [85].

Reform-oriented educational practices highlight the efficacy of learner-centered settings where students are entrusted with greater autonomy and responsibility for their own learning [10]. As delineated in [85], such an educational setting attends and responds to the knowledge, abilities, perspectives, and attitudes of learners. Moreover, an educator in such a setting begins by diagnosing the prevailing status of what the learners know and what they can do. Finally, engaging students in activities that cater to their experiences and interests engenders culturally responsive learning wherein students build their own knowledge. In the spirit of [85], educational robotics embodies the promise of a learner-centered setting whose realization mandates thoughtful incorporation of robotics for teaching and learning. In a robotics-based educational setting, an array of techniques can be employed to determine the status of learner knowledge and abilities [67]. Furthermore, students can be engaged in actively creating their robotics devices [64], constructing their own knowledge and understanding, and practicing new skills. See Figures 1.6 (a)–(b) illustrating instructors responding to student needs during robotics explorations.

The mandate of a knowledge-centered learning setting is to identify students' preconceptions, support their disciplinary understanding, and engage them in sense-making [85]. Educational robotics can be deployed to create a knowledge-centered learning setting wherein student preconceptions are identified using the tools suggested in [67]. Moreover, as students perform hands-on activities and exchange ideas, their conceptual understanding is further revealed. To support students' disciplinary understanding, myriad standards-aligned science and math lessons are suggested in literature that employ experiential learning with robotics [29, 45, 71]. Finally, actively experimenting with a robot and observing its outcomes inherently engender sense-making and metacognition as one seeks to understand the operation, behavior, or state of the robot [66].

(a)

(b)

FIGURES 1.6 (a)–(b) Instructors respond to learner queries during educational robotics explorations to render a learner-centered environment.

An assessment-centered learning setting must proactively interweave assessment opportunities to: review student work, provide them feedback, and allow them to revise their work. A robotics-based learning setting can embed varied assessment strategies [67], including observation, content test, design and code review, experimental demonstration, oral presentation, and written report, among others. Many of the suggested strategies permit review, feedback, and revision tasks suggested above. While some of these approaches entail formative assessment to enhance teaching and learning, others support summative assessment to gauge student progress [67].

Finally, a community-centered learning setting entails formulating and operationalizing beliefs, customs, and practices concerning classroom, schools, and broader community [85] that catalyze student learning. The incorporation of robotics in an educational setting can engender collaborative exploration and dialogue with peers in a classroom. Moreover, students may be provided opportunities to demonstrate their robotics creations to their school community at a showcase or exhibition [54]. In a robotics-based educational setting, learners may be tasked to formulate and address real-world problems in their own communities [35, 86]. Such an approach invites learners to interface with their community, draw from their funds of knowledge [87], grow agency to mold the product of their work [88], and be empowered to solve their community's problems.

(x) Triggers myriad person-level factors

Researchers from varied fields have examined myriad person-level cognitive, emotional, and psychological factors, such as curiosity [89], engagement [90], interest [91], and joy [92], among others, for their role in promoting desired learning behaviors among students. In fact, the use of educational robots in learning environments is often cited to give rise to these and other similar constructs

FIGURES 1.7 (a)–(c) A student examines a robotics artifact with curiosity (a), a student group partakes in a robotics activity with interest and engagement (b), and two visitors to an exhibit look at a robot with joy (c).

(see Figures 1.7 (a)–(c)). For example, students' innate fascination with robots offers a compelling hook [93] to evoke their curiosity. As students operate a robot, explore its behavior, or sense its environment, their curiosity is sparked to probe about the underlying STEM principles and tools that drive the observed phenomena. In contrast to learning interfaces that employ interactive screens, educational robots are deemed to engender superior engagement due to their physical embodiment [47, 94]. Robots have additionally been shown to enhance student interest in STEM disciplines through interactive learning experiences [47, 95]. The use of educational robotics cultivates among students a sense of agency and an ownership of the learning task, which in turn promote their engagement and interest [10]. Thoughtful integration of robots in addressing real-world challenges that have societal relevance and hold personal meaning for students can catalyze their interest in STEM learning. Given a choice, people will often elect to perform activities that they consider fun and enjoyable. When students are given an opportunity to perform hands-on manipulations with a robotics artifact during classroom learning, they find such active learning meaningful and the resulting experience enjoyable [81]. Moreover, when conducting a robot-based learning activity, recall of joyful childhood play with construction blocks [42] can transform the new experience into a fun memory.

4. Exhibiting the role of robots in supporting student learning

A literature review on the applications of educational robotics by Mubin et al. [65] has classified three categories through which robotics can be embedded in learning activities: (1) as a learning tool where programming and building robots using sensors and actuators can enable students to learn about a topic; (2) as a peer that can render spontaneous feedback and interaction with a robot to learn, for example, a robot and student collaboratively solving problems [96]; and (3) as a mentor where the robot adapts to a student's learning level as they solve problems [97]. Common programs that use robotics as a technological tool to enhance instruction typically involve students in building robots, programming them, and analyzing their behaviors within their operating environment [72]. Use of robotics in K-12 education is frequently limited to robot design and programming even though it offers myriad opportunities to deepen understanding of various other concepts [45] that students learn during their K-12 education. When robotics is used for instruction as a passive device it limits the scope of its application [47] and fails to exploit the fullest potential of this powerful learning tool. There are a plethora of ways in which robots can be used as active and constructionist tools to promote student interest and engagement in formal learning in classrooms and laboratories, informal learning through summer camps, and project-based content-specific implementations, etc. To understand the role of a robot as a learning tool, a study [47] sought to examine its ability in scaffolding student learning and observe the challenges that affect it. Six teachers, who had previously undergone training and received implementation support to effectively integrate robotics in STEM education, created various robotics-based lessons and learning plans for classroom implementation. The plans were presented to a group of students, peers, and researchers and were observed and analyzed to study the role of robots in these lessons. The description of one illustrative lesson and findings for it are summarized below.

4.a. Description of the robotics learning artifact

LEGO Mindstorms EV3 kits have been used to implement the lesson in this case study. LEGO robotics kits consist of hardware and software components that are used to create fully functioning robotics systems valuable for teaching and learning [98]. The creation of these constructionist tools was co-conceptualized by Seymour Papert, developer of the theory of constructionism [6], and Kjeld Kirk Kristiansen, then Chief Executive Officer (CEO) of The LEGO Group [98]. Since its inception, LEGO robotics kits have undergone several iterations and have emerged as the most popular robotics tool used for educational purposes [98]. According to Gura [99], LEGO robotics "is a body of teaching and learning practice based on LEGO Robotics kits, popular sets of materials that enable individuals without formal training in engineering and computer programming to design, build, and program small-scale robots."

LEGO robotics kits consist of several parts (for instance, LEGO Mindstorms EV3 consists of over 600 components) including motors, gears, wheels, and other mechanical elements to move the robot, modular LEGO parts to construct the body, and sensors to identify and quantify conditions in the robot's vicinity. The core component of these kits is a programmable brick that is a small processor used to provide commands to move the robot and to perform desired tasks [99]. While this programmable brick has the capacity to run a limited set of commands, most of the programming is done on a computer, using the designated LEGO Mindstorms EV3 software, and then transferred to the brick [99]. In a typical learning setup, students build their robot's body using the construction pieces and use computer software to program the brick of their robot [99]. With the help of their teacher, students keep modifying their design till they reach their intended goal, and work on solving problems by manipulating the data and synthesizing their learning [99].

Note: There are myriad other robotics kits (e.g., Parallax BoeBot [100], Edison Robots [101]) available for learners and the one that aligns most closely with the intended learning goal and audience should be procured by the educators.

4.b. Brief narrative of the lesson [47]

One of the teachers prepared and implemented a lesson on an eighth-grade topic of 'a system of linear equations.' The lesson utilized a project-based learning approach to foster 21st-century skills and to give opportunities to impart soft skills alongside math learning. First, the robots were used to explain the topics of proportionality and rate of change. This was done by programming robots to move along the lines with different slopes (as depicted in Figure 1.8). Students used different robots at 10 student stations and operated them at various speeds to understand the notion of proportionality and rate of change. Through varied setups, students measured the time traveled and distance covered by the robot and plotted a graph. From their experimentation and calculations of speed, distance, and time, students explored the formulation of the direct variation equation $y=kx$.

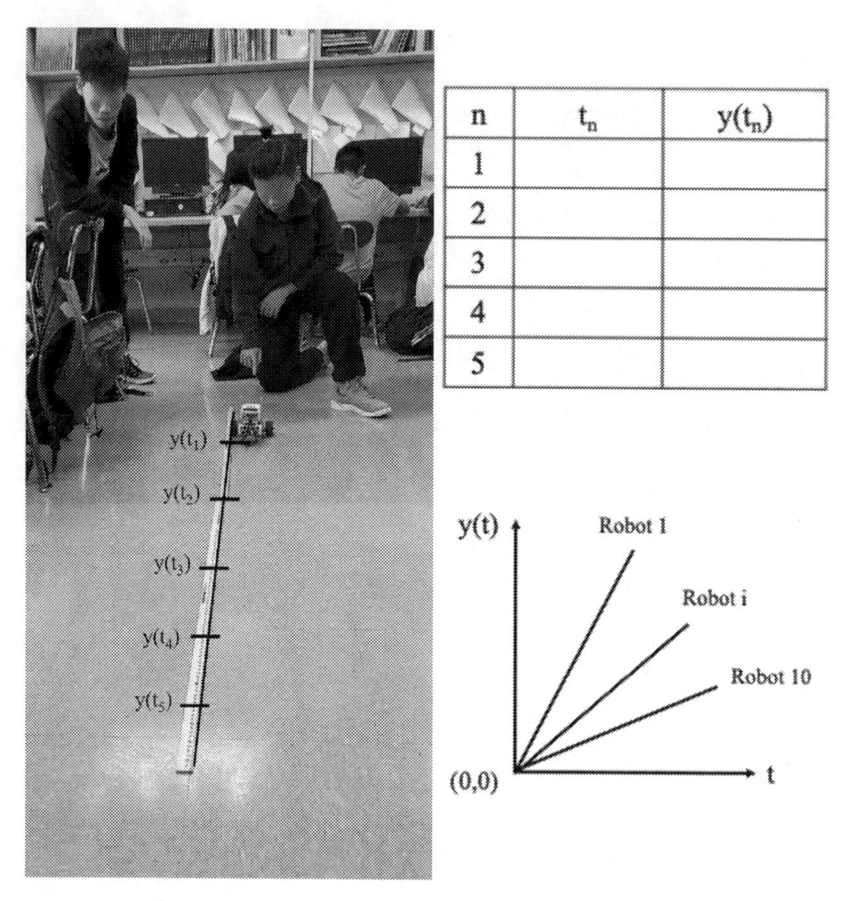

FIGURE 1.8 Schematic representation of a lesson on 'rate of change.'

Students' understanding was broadened from the equation of a line to a system of linear equations, and three methods of finding the solution for two equations, that is, elimination, substitution, and graphing were explored. For the final project, using the collaborative learning approach [102], students formulated real-world problems from their personal experiences with storylines that could be reasonably approximated by a system of two linear equations and worked on solving them in groups (as depicted in Figures 1.9 (a)–(b)). Some examples of storylines included comparison between the quantity of smartphones manufactured by two prominent brands in a specified duration, or time taken to poll the same number of votes by two electoral candidates, among others. The storylines were translated into an appropriate system of linear equations. To solve the problems, the students coded the robots to traverse along two straight lines corresponding to each constraint and observed their intersection point. Later, they tested the graphical solution using the other two methods of elimination and substitution and presented their findings to their peers.

(a)

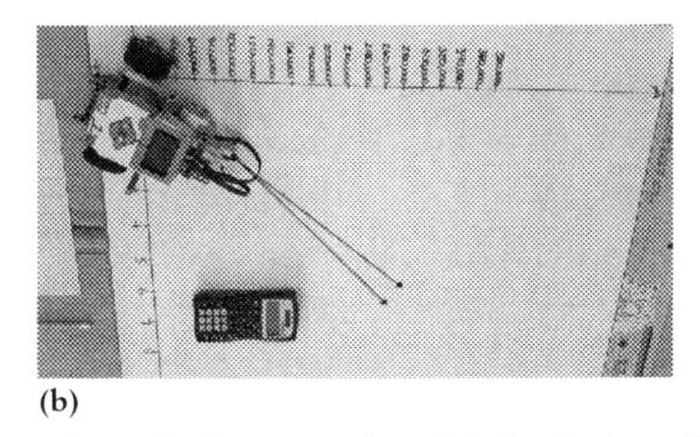

(b)

FIGURES 1.9 (a)–(b) Schematic representation of a lesson on 'graphical solution of a system of linear equations.' Students draw two graphs (a) to represent two linear equations whose slopes and y-axis intercepts are specified. They update their robot code with the slope and y-axis intercept information. For each linear equation, they place the robot at the corresponding y-axis intercept, execute the code, and verify that the robot moves along the graph (b) and passes through the intersection point for the two graphs.

4.c. Findings from the lesson implementation [47]

The teacher believed that involving students with real-life situations keeps them engaged in learning and acts as a medium to express their interests. For instance, it was seen that one student's storyline revolved around submarines as she aspired to be a Navy SEAL. Furthermore, robotics-enabled activities helped the learners to visualize the two-dimensional concepts of space, position, and directionality, and enhanced their reasoning skills. The activities also enriched language development, collaboration, presentation, and research skills among students. The lesson helped the learners in translating their manual representations to symbolic representations, and in graphically solving the system of equations. These solution-finding techniques rendered opportunities to understand the behavior of linear systems, to build upon learners' prior knowledge, and to visually access abstract information.

The findings from this particular lesson case study, and others in [47], reveal certain principal roles of LEGO robots in scaffolding student learning such as: developing learners' creativity and engineering design, fostering opportunities to understand and absorb information visually, and improving learners' collaboration and communication skills. In line with findings from this study, [72] has also evidenced that the use of robots as artifacts illustrates an overall positive learning attitude and a strong preference for hands-on learning among students. Moreover, students acknowledged the improvement in learning effectiveness and engagement with the use of LEGO robots as artifacts in their classrooms [72].

4.d. General challenges in implementing robotics-enabled lessons

While integrating robotics in the classroom can increase enthusiasm, interest, and engagement in learning, it also poses some common challenges. The study of [47] revealed that robots as tools lack built-in mechanisms that could guide individual students in making progress, that is, as a stand-alone tool the robot may not have sufficient capability to support learners in finding logical solutions. For example, as revealed by the reflections of the study participants, when used as a learning tool by novice students, robots may not be able to provide scaffolding to them. Furthermore, when planning instruction, teachers may use robots to only perform procedural tasks as opposed to higher-order cognitive tasks, thus limiting the full potential of using such devices in classrooms [47]. As mentioned earlier, effective technology integrations do not merely use technology as a direct substitution tool without any functional improvement but significantly redefine the function to build authentic and engaging experiences [32]. This requires a foundational understanding of the device functionalities and instructional methods for an effective integration.

Logistically, while using robots or any technology in the classroom, a sufficient number of devices must be present [45] to ensure that students get an adequate amount of hands-on learning time. Enough number of devices ensure that the students are engaged and get opportunities to accomplish the goals of the lesson. But financial limitations might constrain the number of devices that can be made available [72]. Moreover, sensors provided with these robots are not high-precision scientific instruments and sometimes may cause reliability errors in data collection [47]. The following chapters in this book suggest methods to ease the challenges and enable a seamless integration of educational robotics for classroom learning.

5. Conclusion

Educational robotics has seen an increasing attention among educators in transforming classroom learning in the past few years [5]. It was originally introduced to increase student interest and engage them in topics that are challenging to comprehend via conventional methods [8]. Subsequent research has revealed multiple affordances of using robotics such as: quick adaptability across age-levels [45, 53, 56, 71]; cognitive and social development [4]; teaching multiple disciplines [35]; instilling positive learning attitude [72]; exploring advanced content knowledge through hands-on activities [35]; giving opportunities to collaborate, think critically, and problem-solve [35]; fostering 21st century skills [6]; contributing to bridge the STEM learning gap [4]; engendering diverse and engaging learning opportunities [72]; among others. Although educational robotics is a promising tool to enhance interest and effectiveness, some challenges may constrain its implementation. In the absence of sufficient exposure and training, teachers may use robots to perform passive procedural tasks as opposed to active higher-order cognitive tasks. This may limit the promise of the robotics tool to facilitate deeper understanding [47]. It has also been noted that robots lack built-in mechanisms to guide students in making individual progress and thus may be insufficient as stand-alone learning tools [47]. Moreover, to ensure ample hands-on time, an adequate number of devices must be present but financial constraints may be a limitation to make such provisions [72]. Overall, overcoming the barriers with adequate teacher preparation and curriculum revitalization can be a gateway for learners to deeply explore and learn abstract STEM topics that often remain inaccessible to them.

6. Key takeaways

- STEM stands for science, technology, engineering, and mathematics and is most meaningful when all the four subject areas are taught as an integrated whole [11]. STEM learning can engender deeper student learning and foster 21st-century skills [17]. Its primary attributes include: hands-on, real-life, and interdisciplinary learning that aid students to enhance their critical thinking, problem-solving, and analytical skills [1].
- Technology-enabled learning creates authentic learning experiences, encourages students in creating their own knowledge, supports them in developing skills such as self-regulation and problem-solving, and helps them in reducing cognitive load [4].
- The field that utilizes robotics to serve educational goals is called educational robotics. Robotics, as a learning tool, can facilitate hands-on activities, give opportunities to collaborate, think critically and problem-solve, and improve learning outcomes in an engaging manner [37]. Robotics is a motivating learning tool since it engenders excitement [66] and enthusiasm [67] among learners as a result of hands-on explorations and ease of use for learners of various age groups.
- Various robotics components including hardware, such as sensors and construction blocks, and software, such as coding applications, afford opportunities to prototype, design, and explore robots in different situations. This allows students to creatively explore solutions and make sense of abstract topics [8, 9]. Furthermore, it allows teachers to adopt varied pedagogical approaches to achieve learning goals [38, 39].
- LEGO robots, or other similar technologies, provide scaffolding in learning such as developing creativity and engineering design, absorbing information visually, and improving collaboration and communication skills [47]. These tools also instill a positive learning attitude among students and engender diverse and engaging learning opportunities [72], especially those pertaining to STEM learning.
- Effective integration of robotics in learning requires a deep understanding of varied functionalities of robots in alignment with instructional goals [32]. The absence of such an understanding may limit the optimal use of robots as learning tools. Provision of an adequate number of robotics devices and sufficient duration of hands-on exploration are necessary to accomplish the intended learning goals [45].

References

1. Brown, R., Brown, J., Reardon, K., and Merrill, C., Understanding STEM: Current perceptions. *Technology and Engineering Teacher*, 2011.70(6): p. 5–10.
2. Bybee, R.W., *The Case for STEM Education: Challenges and Opportunities*. 2013, Arlington, VA: NSTA Press.
3. Yang, D. and Baldwin, S.J., Using technology to support student learning in an integrated STEM learning environment. *International Journal of Technology in Education and Science*, 2020.4(1): p. 1–11.
4. Ottenbreit-Leftwich, A.T., Glazewski, K.D., Newby, T.J., and Ertmer, P.A., Teacher value beliefs associated with using technology: Addressing professional and student needs. *Computers & Education*, 2010.55(3): p. 1321–1335.
5. Alimisis, D., Educational robotics: Open questions and new challenges. *Themes in Science and Technology Education*, 2013.6(1): p. 63–71.
6. Ackermann, E., Piaget's constructivism, Papert's constructionism: What's the difference? in *Proceedings of Constructivism: Uses and Perspectives in Education*. 2001, Geneva, Switzerland: Research Center in Education. p. 85–94; Available from: https://learning.media.mit.edu/content/publications/EA.Piaget%20_%20Papert.pdf.
7. Blancas, M., Valero, C., Vouloutsi, V., Mura, A., and Verschure, P.F., Educational robotics: A journey, not a destination, in *Handbook of Research on Using Educational Robotics to Facilitate Student Learning*, S. Papadakis and M. Kalogiannakis, Editors. 2021, Hershey, PA: IGI Global. p. 41–67.
8. Eguchi, A., Educational robotics theories and practice: Tips for how to do it right, in *Robots in K-12 Education: A New Technology for Learning*, B.S. Barker, Editor. 2012, Hershey, PA: IGI Global. p. 1–30.
9. Eguchi, A., Theories and practices behind educational robotics for all, in *Handbook of Research on Using Educational Robotics to Facilitate Student Learning*. 2021, Hershey, PA: IGI Global. p. 68–106.
10. Sabouri, P., Ghosh, S., Mallik, A., and Kapila, V., The formation and dynamics of teacher roles in a teacher-student groupwork during a robotic project, in *Proceedings of ASEE Annual Conference and Exposition*. 2020, Virtual; Available from: https://peer.asee.org/35323.
11. Dugger, W.E., Evolution of STEM in the United States, in *Proceedings of International Conference on Technology Education Research*. December 2010, Gold Coast, Queensland, Australia.
12. Merrill, C. and Daugherty, J. The future of TE masters degrees: STEM, in *Proceedings of International Technology Education Association Conference*. 2009, Louisville, KY.
13. National Science Foundation, *About NSF*; Available from: https://www.nsf.gov/about/glance.jsp.
14. White, D.W., What is STEM education and why is it important? *Florida Association of Teacher Educators Journal*, 2014.1(14): p. 1–9.
15. National Science Board, *A National Action Plan for Addressing the Critical Needs of the US Science, Technology, Engineering, and Mathematics Education System*. 2007, Arlington, VA: National Science Foundation.
16. Moore, T.J., Stohlmann, M.S., Wang, H.-H., Tank, K.M., Glancy, A.W., and Roehrig, G.H., Implementation and integration of engineering in K-12 STEM education, in *Engineering in Pre-College Settings: Synthesizing Research, Policy, and Practices*, Ş. Purzer, J. Strobel, and M.E. Cardella, Editors. 2014, West Lafayette: Purdue University Press. p. 35–60.

17. Honey, M., Pearson, G., and Schweingruber, H.A., *STEM Integration in K-12 Education: Status, Prospects, and an Agenda for Research*. 2014, Washington, DC: National Academies Press.
18. Thahir, A., Anwar, C., Saregar, A., Choiriah, L., Susanti, F., and Pricilia, A., The effectiveness of STEM learning: Scientific attitudes and students' conceptual understanding. *Journal of Physics: Conference Series*, 2020.1467(1): p. 012008.
19. Morado, M.F., Melo, A.E., and Jarman, A., Learning by making: A framework to revisit practices in a constructionist learning environment. *British Journal of Educational Technology*, 2021.52(3): p. 1093–1115.
20. Kafai, Y.B., Constructionism, in *The Cambridge Handbook of the Learning Sciences*, R.K. Sawyer, Editor. 2005, Cambridge, UK: Cambridge University Press. p. 35–46.
21. Bevan, B., Ryoo, J., and Shea, M., What if? Building creative cultures for STEM making and learning. *Afterschool Matters*, 2017.25: p. 1–8.
22. Benitti, F.B.V., Exploring the educational potential of robotics in schools: A systematic review. *Computers & Education*, 2012.58(3): p. 978–988.
23. Riojas, M., Lysecky, S., and Rozenblit, J., Educational technologies for precollege engineering education. *IEEE Transactions on Learning Technologies*, 2011.5(1): p. 20–37.
24. Young, J.L., Young, J.R., and Ford, D.Y., Culturally relevant STEM out-of-school time: A rationale to support gifted girls of color. *Roeper Review*, 2019.41(1): p. 8–19.
25. Chen, F.H., Looi, C.K., and Chen, W., Integrating technology in the classroom: A visual conceptualization of teachers' knowledge, goals and beliefs. *Journal of Computer Assisted Learning*, 2009.25(5): p. 470–488.
26. Ottenbreit-Leftwich, A., Liao, J.Y.-C., Sadik, O., and Ertmer, P., Evolution of teachers' technology integration knowledge, beliefs, and practices: How can we support beginning teachers use of technology? *Journal of Research on Technology in Education*, 2018.50(4): p. 282–304.
27. AeA, *Losing the Competitive Advantage? The Challenge for Science and Technology in the United States*. 2005, Washington, DC: American Electronics Association.
28. Symonds, W.C., Schwartz, R., and Ferguson, R.F., *Pathways to Prosperity: Meeting the Challenge of Preparing Young Americans for the 21st Century*. 2011, Cambridge, MA: Harvard University Graduate School of Education.
29. Ghosh, S., Krishnan, V.J., Rajguru, S.B., and Kapila, V., Middle school teacher professional development in creating a NGSS-plus-5E robotics curriculum (Fundamental), in *Proceedings of ASEE Annual Conference and Exposition*. 2019, Tampa, FL; Available from: https://peer.asee.org/33108.
30. Brill, A.S., Elliott, C.H., Listman, J.B., Milne, C., and Kapila, V., Middle school teachers' evolution of TPACK understanding through professional development, in *Proceedings of ASEE Annual Conference and Exposition*. 2016, New Orleans, LA; Available from: https://peer.asee.org/25720.
31. Brill, A.S., Listman, J.B., and Kapila, V., Using robotics as the technological foundation for the TPACK framework in K-12 classrooms, in *Proceedings of ASEE Annual Conference and Exposition*. 2015, Seattle, WA; Available from: https://peer.asee.org/25015.
32. Puentedura, R.R., *Transformation, Technology, and Education*. 2006; Available from: http://hippasus.com/resources/tte/.
33. Frank, J.A. and Kapila, V., Mixed-reality learning environments: Integrating mobile interfaces with laboratory test-beds. *Computers & Education*, 2017.110: p. 88–104.

34. Mallik, A. and Kapila, V., Interactive learning of mobile robots kinematics using ARCore, in *Proceedings of IEEE International Conference on Robotics and Automation Engineering (ICRAE)*. 2020. p. 1–6.
35. Mallik, A., Liu, D., and Kapila, V., Analyzing the outcomes of a robotics workshop on the self-efficacy, familiarity, and content knowledge of participants and examining their designs for end-of-year robotics contests. *Education and Information Technologies*. 2022; Available from: https://doi.org/10.1007/s10639-022-11400-1.
36. Angelopoulos, P., Mitropoulou, D., and Papadimas, K., The contribution of open educational robotics competition to support STEM education, in *Handbook of Research on Using Educational Robotics to Facilitate Student Learning*. 2021, Hershey, PA: IGI Global. p. 41–67.
37. Eguchi, A., Robotics as a learning tool for educational transformation, in *Proceedings of International Workshop Teaching Robotics, Teaching with Robotics and International Conference Robotics in Education*. 2014, Padova, Italy. p. 27–34; Available from: https://www.terecop.eu/TRTWR-RIE2014/files/00_WFr1/00_WFr1_04.pdf.
38. Elmore, B.B. and Seiler, E., Using LEGO™ robotics for K-12 engineering outreach, in *Proceedings of ASEE Southeast Section Conference*. 2008, Memphis, TN. p. 1–5; Available from: http://se.asee.org/proceedings/ASEE2008/papers/RP2008007ELM.pdf.
39. Nekovei, R. and Nekovei, D., Teaching by design: An early introduction to science, technology, engineering and mathematics (STEM) concepts, in *Proceedings of International Conference on Engineering Education*. 2004, Gainesville, FL. p. 1–4; Available from: https://www.ineer.org/Events/ICEE2004/Proceedings/Papers/301_NekoveiICEE_(1).pdf.
40. Mallik, A., Sabouri, P., Ghosh, S., and Kapila, V., Assessing the effects of a robotics workshop with draw-a-robot test (Fundamental), in *Proceedings of ASEE Annual Conference and Exposition*. 2020, Virtual; Available from: https://peer.asee.org/34182.
41. You, H.S. and Kapila, V., Effectiveness of professional development: Integration of educational robotics into science and math curricula, in *Proceedings of ASEE Annual Conference and Exposition*. 2017, Columbus, OH; Available from: https://peer.asee.org/28207.
42. Whitman, L. and Witherspoon, T., Using LEGOs to interest high school students and improve K12 STEM education, in *Proceedings of ASEE/IEEE Frontiers in Education Conference*. 2003, Boulder, CO. p. F3A6–10; Available from: https://archive.fie-conference.org/fie2003/papers/1366.pdf.
43. Martin, F.G., *Robotics Explorations: A Hands-On Introduction to Engineering*. 2001, Upper Saddle River, NJ: Prentice Hall.
44. VEX, *VEX Robotics Home Page*. 2022; Available from: https://www.vexrobotics.com/.
45. Williams, K., Igel, I., Poveda, R., Kapila, V., and Iskander, M., Enriching K-12 science and mathematics education using LEGOs. *Advances in Engineering Education*, 2012.3(2): p. 1–27.
46. Moorhead, M., Listman, J.B., and Kapila, V., A robotics-focused instructional framework for design-based research in middle school classrooms, in *Proceedings of ASEE Annual Conference and Exposition*. 2015, Seattle, WA; Available from: https://peer.asee.org/23444.
47. Ghosh, S., Sabouri, P., and Kapila, V., Examining the role of LEGO robots as artifacts in STEM classrooms (Fundamental), in *Proceedings of ASEE Annual Conference and Exposition*. 2020, Virtual; Available from: https://peer.asee.org/34620.

48. Collins, A., *Cognitive Apprenticeship: Teaching the Craft of Reading, Writing, and Mathematics. Technical Report No. 403*. 1987, Cambridge, MA: BBN Laboratories, Centre for the Study of Reading, University of Illinois.
49. Chambers, J.M., Carbonaro, M., Rex, M., and Grove, S., Scaffolding knowledge construction through robotic technology: A middle school case study. *Electronic Journal for the Integration of Technology in Education*, 2007.6: p. 55–70.
50. Scott, P., Asoko, H., and Leach, J., Student conceptions and conceptual learning in science, in *Handbook of Research on Science Education*, S. Abell and N. Lederman, Editors. 2013, New York, NY: Routledge. p. 31–56.
51. Reiser, B.J., Scaffolding complex learning: The mechanisms of structuring and problematizing student work. *The Journal of the Learning sciences*, 2004.13: p. 273–304.
52. Davis, E.A. and Miyake, N., Explorations of scaffolding in complex classroom systems. *The Journal of the Learning Sciences*, 2004.13: p. 265–272.
53. Krishnamoorthy, S.P. and Kapila, V., Using a visual programming environment and custom robots to learn c programming and K-12 STEM concepts, in *Proceedings of ACM Annual Conference on Creativity and Fabrication in Education*. 2016, Stanford, CA. p. 41–48.
54. Rusk, N., Resnick, M., Berg, R., and Pezalla-Granlund, M., New pathways into robotics: Strategies for broadening participation. *Journal of Science Education and Technology*, 2008.17(1): p. 59–69.
55. Chen, N.-S., Quadir, B., and Teng, D.C., Integrating book, digital content and robot for enhancing elementary school students' learning of English. *Australasian Journal of Educational Technology*, 2011.27(3): p. 546–561.
56. Rogers, C. and Portsmore, M., Bringing engineering to elementary school. *Journal of STEM Education: Innovations and Research*, 2004.5(3–4): p. 17–28.
57. Bers, M.U., *Blocks to Robots: Learning with Technology in the Early Childhood Classroom*. 2008, New York, NY: Teachers College Press.
58. Björling, E.A. and Rose, E., Participatory research principles in human-centered design: Engaging teens in the co-design of a social robot. *Multimodal Technologies and Interaction*, 2019.3(1): p. 8.
59. Fasola, J. and Matarić, M.J., A socially assistive robot exercise coach for the elderly. *Journal of Human-Robot Interaction*, 2013.2(2): p. 3–32.
60. Ryan, R.M. and Deci, E.L., Intrinsic and extrinsic motivations: Classic definitions and new directions. *Contemporary Educational Psychology*, 2000.25(1): p. 54–67.
61. Moorhead, M., Elliott, C.H., Listman, J.B., Milne, C.E., and Kapila, V., Professional development through situated learning techniques adapted with design-based research, in *Proceedings of ASEE Annual Conference and Exposition*. 2016, New Orleans, LA; Available from: https://peer.asee.org/25967.
62. Petre, M. and Price, B., Using robotics to motivate 'back door' learning. *Education and Information Technologies*, 2004.9(2): p. 147–158.
63. Arís, N. and Orcos, L., Educational robotics in the stage of secondary education: Empirical study on motivation and STEM skills. *Education Sciences*, 2019.9(2): p. 73.
64. Yuen, T., Boecking, M., Stone, J., Tiger, E.P., Gomez, A., Guillen, A., and Arreguin, A., Group tasks, activities, dynamics, and interactions in collaborative robotics projects with elementary and middle school children. *Journal of STEM Education*, 2014.15(1): p. 39–45.

65. Mubin, O., Stevens, C.J., Shahid, S., Al Mahmud, A., and Dong, J.-J., A review of the applicability of robots in education. *Journal of Technology in Education and Learning*, 2013.1: p. 1–7.
66. Faisal, A., Kapila, V., and Iskander, M.G., Using robotics to promote learning in elementary grades, in *Proceedings of ASEE Annual Conference and Exposition*. 2012, San Antonio, TX; Available from: https://peer.asee.org/22196.
67. Krishnan, V.J., Rajguru, S.B., and Kapila, V., Analyzing successful teaching practices in middle school science and math classrooms when using robotics (Fundamental), in *Proceedings of ASEE Annual Conference and Exposition*. 2019, Tampa, FL; Available from: https://peer.asee.org/32092.
68. Ruiz-del-Solar, J. and Avilés, R., Robotics courses for children as a motivation tool: The Chilean experience. *IEEE Transactions on Education*, 2004.47(4): p. 474–480.
69. Hartmann, S., Wiesner, H., and Wiesner-Steiner, A., Robotics and gender: The use of robotics for the empowerment of girls in the classroom, in *Gender Designs IT*, I. Zorn, S. Maass, E. Rommes, C. Schirmer, and H. Schelhowe, Editors. 2007, Berlin: Springer. p. 175–188.
70. Milto, E., Rogers, C., and Portsmore, M., Gender differences in confidence levels, group interactions, and feelings about competition in an introductory robotics course, in *Proceedings of Annual ASEE/IEEE Frontiers in Education Conference*. 2002, Boston, MA. p. F4C7–F4C14.
71. Muldoon, J., Phamduy, P.T., Le Grand, R., Kapila, V., and Iskander, M.G., Connecting cognitive domains of Bloom's taxonomy and robotics to promote learning in K-12 environment, in *Proceedings of ASEE Annual Conference and Exposition*. 2013, Atlanta, GA; Available from: https://peer.asee.org/19343.
72. Williams, K., Kapila, V., and Iskander, M.G., Enriching K-12 science education using LEGOs, in *Proceedings of ASEE Annual Conference and Exposition*. 2011, Vancouver, Canada; Available from: https://peer.asee.org/17911.
73. Hansen, D.M., Larson, R.W., and Dworkin, J.B., What adolescents learn in organized youth activities: A survey of self-reported developmental experiences. *Journal of Research on Adolescence*, 2003.13(1): p. 25–55.
74. Dede, C., *21st Century Skills: Rethinking How Students Learn*, J.A. Bellanca and R. Brandt, Editors. 2010, Bloomington, IN: Solution Tree Press. p. 51–76.
75. Trilling, B. and Fadel, C., *21st Century Skills: Learning for Life in Our Times*. 2009, San Francisco, CA: Jossey-Bass.
76. Greenhill, V., *21st Century Knowledge and Skills in Educator Preparation*. 2010, American Association of Colleges of Teacher Education and Partnership for 21st Century Skills.
77. Rahman, S.M.M., Chacko, S.M., Rajguru, S.B., and Kapila, V., Fundamental: Determining prerequisites for middle school students to participate in robotics-based STEM lessons: A computational thinking approach, in *Proceedings of ASEE Annual Conference and Exposition*. 2018, Salt Lake City, UT; Available from: https://peer.asee.org/30549.
78. Wing, J.M., Computational thinking. *Communications of the ACM*, 2006.49(3): p. 33–35.
79. You, H.S., Chacko, S., and Kapila, V., Teaching science with technology: Scientific and engineering practices of middle school science teachers engaged in a robot-integrated professional development program (Fundamental), in *Proceedings of ASEE Annual Conference and Exposition*. 2019, Tampa, FL; Available from: https://peer.asee.org/33353.

80. Nugent, G., Barker, B., and Grandgenett, N., The impact of educational robotics on student STEM learning, attitudes, and workplace skills, in *Robots in K-12 Education: A New Technology for Learning*. 2012, Hershey, PA: IGI Global.
81. Karim, M.E., Lemaignan, S., and Mondada, F. A review: Can robots reshape K-12 STEM education? in *Proceedings of IEEE International Workshop on Advanced Robotics and its Social Impacts (ARSO)*. 2015, Lyon, France. p. 1–8.
82. You, H.S., Chacko, S.M., Rajguru, S.B., and Kapila, V., Designing robotics-based science lessons aligned with the three dimensions of NGSS-plus-5E model: A content analysis (Fundamental), in *Proceedings of ASEE Annual Conference and Exposition*. 2019, Tampa, FL; Available from: https://peer.asee.org/32622.
83. NGSS, *Next Generation Science Standards (NGSS): For States, By States*. 2013, Washington, DC: The National Academies Press; Available from: https://www.nextgenscience.org/.
84. Tucker-Raymond, E., Varelas, M., Pappas, C.C., Korzh, A., and Wentland, A., "They probably aren't named Rachel": Young children's scientist identities as emergent multimodal narratives. *Cultural Studies of Science Education*, 2007.1(3): p. 559–592.
85. National Research Council, *How People Learn: Brain, Mind, Experience, and School: Expanded Edition*. 2000, Washington, DC: National Academies Press.
86. Chauhan, P. and Kapila, V., Promoting engineering research with entrepreneurship and industry experiences: A teacher professional development program, in *International Encyclopedia of Education, 4th Edition*, R. Tierney, F. Rizvi, and K. Ercikan, Editors. 2022, Amsterdam: Elsevier Science. 11: p. 298–311.
87. González, N., Moll, L.C., and Amanti, C., *Funds of Knowledge: Theorizing Practices in Households, Communities, and Classrooms*. 2006, New York, NY: Routledge.
88. Calabrese Barton, A. and Tan, E., A longitudinal study of equity-oriented STEM-rich making among youth from historically marginalized communities. *American Educational Research Journal*, 2018.55(4): p. 761–800.
89. Reio Jr, T.G., Petrosko, J.M., Wiswell, A.K., and Thongsukmag, J., The measurement and conceptualization of curiosity. *The Journal of Genetic Psychology*, 2006.167(2): p. 117–135.
90. Wong, Z.Y. and Liem, G.A.D., Student engagement: Current state of the construct, conceptual refinement, and future research directions. *Educational Psychology Review*, 2021.34: p. 107–138.
91. Krapp, A., *The Construct of Interest: Characteristics of Individual Interests and Interest-Related Actions from the Perspective of a Person-Object-Theory*. 1993, Munich, Germany Inst. für Erziehungswiss. Und Pädag. Psychologie, Univ. der Bundeswehr.
92. Johnson, M.K., Joy: A review of the literature and suggestions for future directions. *The Journal of Positive Psychology*, 2020.15(1): p. 5–24.
93. Laut, J., Kapila, V., and Iskander, M.G., Exposing middle school students to robotics and engineering through LEGO and MATLAB, in *Proceedings of ASEE Annual Conference and Exposition*. 2013, Atlanta, GA; Available from: https://peer.asee.org/19597.
94. McNerney, T.S., From turtles to tangible programming bricks: Explorations in physical language design. *Personal and Ubiquitous Computing*, 2004.8(5): p. 326–337.
95. Toh, L.P.E., Causo, A., Tzuo, P.-W., Chen, I.-M., and Yeo, S.H., A review on the use of robots in education and young children. *Journal of Educational Technology & Society*, 2016.19(2): p. 148–163.

96. Janssen, J.B., Wal, C.C., Neerincx, M.A., and Looije, R., Motivating children to learn arithmetic with an adaptive robot game, in *Proceedings of International Conference on Social Robotics*. 2011, Berlin: Springer-Verlag. p. 153–162.
97. Hashimoto, T., Kobayashi, H., Polishuk, A., and Verner, I., Elementary science lesson delivered by robot, in *Proceedings of ACM/IEEE International Conference on Human-Robot Interaction (HRI)*. 2013, Tokyo, Japan. p. 133–134.
98. ÜÇGÜL, M., History and educational potential of LEGO Mindstorms NXT. *Mersin Üniversitesi Eğitim Fakültesi Dergisi*, 2013.9.
99. Gura, M., *Getting Started with LEGO Robotics: A Guide for K-12 Educators [Paperback]*. 2011, Eugene, OR: ISTE.
100. Parallax Boe-Bot Robot; Available from: https://www.parallax.com/boe-bot-robot/.
101. Edison Robots; Available from: https://meetedison.com.
102. Altin, H. and Pedaste, M., Learning approaches to applying robotics in science education. *Journal of Baltic Science Education*, 2013.12(3): p. 365–377.

2

APPLICATIONS OF ROBOTS IN EDUCATIONAL SETTINGS

1. Introduction

Educational robotics is being increasingly recognized as a learning tool that promotes hands-on exploration [1] of abstract topics, fosters collaboration [2], enhances problem-solving skills [3], and creates motivating learning environments [4–6]. The theoretical basis of learning through robotics is primarily grounded in Jean Piaget's *constructivism* [7], which theorizes that students actively construct their own knowledge and understanding through meaning-making experiences, and Seymour Papert's theory of *constructionism* [8], which posits that knowledge is created through active engagement with and physical manipulation of learning artifacts. Thus, in the spirit of constructionist principles, educational robotics allows students to build knowledge by designing, testing, and refining their own ideas as they pursue hands-on, experiential explorations situated in real-world settings [9] and deepen their understanding by applying the newly acquired knowledge to varied contexts [10].

In an educational robotics setting, infusing active learning experiences with personal meaning and relevance for students necessitates the adoption of pedagogical techniques such as project-, competition-, and inquiry-based learning, among others [11]. These instructional approaches align with constructivist/constructionist learning theories and engender deeper comprehension of concepts, improved motivation, and enhanced problem-solving skills. Several types of robots can be used to facilitate such active and collaborative learning experiences, for example, brick-based robots, modular kit-based robots, and pre-assembled robots, of which some utilize blocks- or text-based programming while still others are coded using physical blocks. Moreover, robots can be used in formal and informal learning settings, for example, within a school context by creating

DOI: 10.4324/b23177-3

alignment with science, technology, engineering, and mathematics (STEM) curricular standards, through after-school clubs, at summer camps, etc. Robots can also be accorded different roles, such as that of a learning tool, a peer, or a tutor, based on the intended learning outcomes, developmental levels of students, and availability of resources [1, 10]. Recent years have witnessed several examples of applications of robotics for addressing varied learning needs and producing valuable knowledge for future implementations.

This chapter begins by highlighting the fundamental theories and instructional approaches underlying the application of educational robotics. Next, the chapter details the different roles robots can assume and the variety of settings in which robotics can be applied on the basis of the targeted audience, desired learning outcomes, and availability of resources. Lastly, the chapter provides several illustrative examples of successful programs that incorporated robotics for learning.

2. Constructionism and educational robotics

The grounding of educational robotics on several fundamental education theories explains why it is deemed valuable in learning. The two most significant theories that support educational robotics include Jean Piaget's *constructivism* [7] and Seymour Papert's *constructionism* [8]. Both of these learning theories are learner-centered and they emphasize the importance of students creating their own knowledge and understanding through learning experiences. Constructivism contends that for developing their knowledge, students must engage in interactive experiences to build their understanding of the world around them. During such experiences, students actively engage in learning through meaning-making while teachers enable the learning process by focusing on what sense students are making [12]. To facilitate constructivist learning, students must have opportunities to engage in hands-on investigations and interpretations [12]. Constructivism has a positive impact on students' understanding and motivation [2]. Moreover, it has been proposed that knowledge is socially constructed, that is, students learn through interactions with their peers, teachers, and parents, in the form of *social constructivism* [13]. Under a constructivist framework, in learning from their own experiences students begin to appreciate the relevance and meaning of their newly constructed understanding, all of which help them in the retention of knowledge [2].

Constructionism, stemming from constructivism, is learner-driven and focuses on students building their own knowledge through experiences in constructing and interacting with a physical artifact [12]. While constructivism is concerned with knowledge building through mental processes in one's head, constructionism posits that the process of knowledge creation can be scaffolded by engaging in the process of constructing and manipulating a tangible physical artifact [12]. Thus, constructionism emphasizes the presence of "objects to think with," that is, materials existing in students' environment, that can be applied in the context of disciplinary learning, and that can foster investigations of difficult concepts through "bodily engagement" [14]. The process of knowledge creation through tangible experiences occurs when students link their prior knowledge and new learning while interacting with physical objects [15]. In designing a constructionist learning environment, it is imperative to recognize that learning occurs 'by doing' or 'by making,' knowledge is socially constructed, and opportunities to build things that can be shared are critical [16]. Moreover, in constructionist environments, learning occurs through discussions among teachers and students about the design and interactions with learning artifacts, whether physical or digital [15]. Guided by the constructionist principles, a digital tool named Logo programming language was conceptualized and developed to build an understanding of mathematics, computing, and science concepts among students [15]. The theory of constructionism is also consistent with *situated learning* that suggests that knowledge is constructed in the same social context where it occurs and it is accompanied with concrete "objects to think with" [12]. Contextualized experiences with physical manipulatives can help students examine, probe, and investigate an abstract concept tangibly, which helps in enhancing and sustaining their learning [12]. By engaging students in personally appealing explorations, a constructionist environment offers the potential to deepen their learning while allowing them to explore their innate interests with the use of technology [2, 16].

Robots can serve as a relevant and contemporary technological tool, with both hardware and software capabilities, that offers the promise of allowing students to concretely experiment through construction and interaction activities to build their learning [2, 17]. A robotics-based learning process generally requires the following four steps: designing and building the robot, creating a program for the robot, transferring the program to the robot, and executing the program to operate the robot and analyzing its behavior. These steps align with the principles of constructionist environments that include hands-on experiences entailing opportunities to "identify powerful ideas" as well as to design, construct, and test conceptual ideas, including those from other disciplines [17]. Educational robotics can play the role of a cognitive tool as it can aid students in building their problem-solving capabilities by manipulating a physical object and learning through collaboration, discussion, and reflection [17]. Furthermore, educational robotics embeds opportunities for students to learn by receiving instant feedback on their work with manipulatives [2], it allows learning to be situated in real-world contexts to make it personally relatable for students [9], and it facilitates transfer of learning to newer contexts [10]. To illustrate the operationalization of constructionist learning with educational robotics, consider a lesson on center of mass in which a high school physics teacher used LEGO robots to teach an object's equilibrium point [18]. Specifically, the teacher encouraged experiments where students placed the robot's programmable brick in different positions (front, middle, or back) and directed the robot to go up an incline to find the most stable positioning, emulating a vehicle with load going up a ramp. In addition to noting their observations, students engaged in discussions and presented their rationale to their peers. Aligned with constructionist principles, such a learning experience allowed students to conceptualize an abstract topic by using robots as "objects-to-think-with" and constructing their knowledge socially through discussion and reflection. Moreover, in this example, the learning was situated in the context of real-life simulation of preventing a vehicle from toppling while going up a ramp and it is aligned to grade-level disciplinary learning standards. See Chapter 8, Subsection 4.b, for additional details on this center of mass lesson.

2.a. Instructional approaches to facilitate teaching and learning through robotics

To foster constructivist and/or constructionist learning environments, a variety of pedagogical techniques can be applied for achieving desired learning outcomes, for example, project-based learning (PBL), competition-based learning (CBL), and inquiry-based learning (IBL), among others [11]. These different methods are well-aligned with the hands-on nature of educational robotics and they can help organize the instructional activities effectively. Under the PBL framework, students are challenged to examine and address a problem that encapsulates the complexity inherent in a real-world context accurately [19]. As they are drawn to this personally meaningful undertaking and engage with it over an extended duration, strategically scheduled instructional scaffolds enable students to acquire new knowledge and they receive ample opportunities to hone their skills through practice. Teaching and learning activities under the PBL model are often formulated to be constructionist and contribute to enhancements in student learning in myriad ways, such as expanded understanding of concepts, increased ability for problem-solving, enhanced capacity to connect classroom learning to real-world, and improved attitudes toward learning [20]. Embedding PBL with educational robotics can provide students with opportunities to collaboratively engage in authentic problems [21]. Through projects of personal relevance, students carry out investigations and critically analyze errors in their robot design or programming [21]. Upon receiving instant feedback in the form of robot behavior or performance, as students collectively strategize, modify, and implement new solutions on their own they operationalize self-directed learning [21].

Competition-based learning (CBL) constitutes yet another pedagogical approach to constructivist/constructionist learning wherein the controlled environment of a competition is used to prime student motivation for learning in support of achieving some desired learning outcomes [22]. Under the CBL approach, as students collaborate in groups to work on a project focused on an authentic, real-world problem, they are supported to enhance their skills of creativity, organization, and collaboration [22]. Informal robotics-based competitions have emerged as an early driver for the recognition of the value of robotics in educational settings since they often spark learners' initial interest in science and technology disciplines [23]. FIRST LEGO League (FLL) is a popular example of competition-based learning where students participate in learning collectively and purposefully by generating ideas, devising solution strategies, and utilizing technology for hands-on explorations [23]. Participation in FLL has been shown to enhance student interest in science and engineering subjects, foster an environment for STEM learning, embed opportunities for learning multiple creative ways to debug real-world problems, and improve communication skills [23]. Incorporating CBL with educational robotics is deemed to be a promising approach for supporting students to practice and sharpen their disciplinary knowledge [11]. Nonetheless, the promise of CBL with robotics is limited to only a small cohort of students in any classroom due to significant financial and organizational obligations [11]. Yet, the popularity of CBL with robotics reveals that one can take advantage of students' fascination with and interest in robotics by using this as a hook to stimulate them to learn STEM disciplines. That is, robotics can be embedded in the regular school curriculum and classroom to create positive and engaging learning experiences.

Inquiry-based learning (IBL) is a pedagogical approach wherein students wrestle with authentic, real-world challenges to discover concepts by asking questions, constructing explanations, testing solutions, investigating phenomena, making predictions, drawing conclusions, and communicating results [24]. Engaging in IBL represents a departure from rote memorization of isolated facts [25, 26] and a movement toward the acquisition of a deeper understanding of scientific phenomena [24]. IBL facilitates active participation [26], emphasizes problem solving [24], and helps students in making connections between their pre-existing conceptions and new knowledge [24]. In the context of educational robotics, the IBL approach allows students to engage in the recurring cycles of *inquire-connect-investigate-analyze-feedback-improve*, making learning meaningful and relevant. IBL with the support of peer discussions and collaborations can additionally help students to come up with novel ideas, practice critical thinking, and build on their prior conceptions [27].

As evidenced above, IBL entails teachers and students being engaged in myriad practices deemed consistent with Next Generation Science Standards (NGSS) [28] that support a deep understanding of scientific concepts and are based on the *Framework for K-12 Science Education* ("*Framework*") [29]. The NGSS comprise of science and engineering practices that help students to investigate real-world phenomena and build their way of thinking. Next, the *Framework* suggests considering cognitive, social, and physical practices required to enact inquiry science and experiencing inquiry science actively instead of passively. Finally, the Deweyan perspective [30] suggests engaging learners through authentic problem-solving emulating the practices of real-world practitioners that aligns with the science and engineering practices of NGSS [28]. Thus, consistent with the focus of NGSS [28], the *Framework* [29], and the Deweyan perspective [30], it is paramount to acknowledge the importance of creating experiences that help students' understanding of scientific concepts in real-world contexts as well as inducing a conceptual change, and therefore, employ IBL with educational robotics.

Even though IBL tends to be an ill-defined concept in modern educational settings, this book illustrates inquiry applications in a robotics-based learning setting through several exemplars. Operationalizing inquiry through a range of scientific and engineering practices that can be used to build knowledge mitigates simplistic notions of inquiry as a set of linear steps. Such a methodology is founded on the fundamental characteristics of inquiry that allow students to enact practices such as applying theories to make predictions and formulate possible explanations, asking questions, making observations and measurements, collecting and analyzing data, and formulating arguments and developing solutions. In this spirit, robotics-based learning activities of this book entail engaging students and teachers in investigating, making claims for justifiable explanations, and evaluating such claims. Additionally, the methods delineated in this book support teachers to experience and use a broad range of practices that engage them in exploration of nature and quality of evidence and support them to incorporate this approach to inquiry into the curricula that they develop. Often teachers who mentor robotics teams in informal learning setting are unprepared in the technical knowledge and pedagogical applications of robotics to foster inquiry-based STEM learning. Teachers need to understand how robotics can be instrumental in providing hands-on, inquiry-based activities that promote deep learning of the core concepts, principles, and practices of science and math, and the following chapters illustrate models that provide such experiences.

2.b. Types of educational robotics tools and platforms

Several different types of robotics tools and platforms are available for educational purposes from which one can be chosen for use based on student age group, desired learning outcomes, etc. [1]. Through its well-considered incorporation in a learning activity, a robotics tool can render a constructionist setting for learning with project-, competition-, or inquiry-based pedagogical methods or with a thoughtful mix of multiple approaches as desired. Although there is a multiplicity of ways to distinguish the different varieties of robots used for educational purposes, in this subsection, based on the physical design and programming methods, we classify them as: brick-based robots, modular kit-based robots, and pre-assembled robots, of which some utilize blocks- or text-based programming while still others are coded using physical blocks [1, 2, 31]. Note that many education and hobby robots that are interfaced with and programmed using a variety of microcontrollers, such as Arduino, Raspberry Pi, etc., fall under the kit-based and pre-assembled robots that can be programmed using blocks- or text-based programming methods. Finally, note that the following discussion focuses primarily on mobile robotics platforms and does not address manipulator, underwater, or aerial robots.

By brick-based robots, we refer to educational robotics systems that utilize construction components, such as axles, beams, bushings, connectors, frames, hubs, gears, pegs, treads, among others, which readily interconnect to form the structure and mechanism of a robot. Such robotics systems additionally include a variety of plug-and-play sensors and motors that can be effortlessly mounted on the robot using various components that are part of the robot construction system. Lastly, these robotics systems include a hand-held, portable, programmable brick that can be mounted on the robot for processing sensory input and generating output control signals [32]. Often such systems do not entail the use of any tools for mechanical assembly or any breadboarding or soldering for electronics assembly. After fully assembling the structural, locomotion, sensor, motor, and computing components, to provide directives for sensing and controlling the robot behavior, a desktop or laptop computer is used to create a blocks- or text-based program. Next, the resulting program can be transferred to the programmable brick using a serial cable or a Bluetooth wireless connection. Once a program is downloaded to the robot's programmable brick it remains stored on it and can be executed by making appropriate selections using the brick's liquid crystal display (LCD) screen and various navigation buttons [32]. The popular LEGO Mindstorms robotics system permits one to conceive and create a modular mechanical structure of their robot; flexibly impart locomotion to it using wheels, treads, or legs; and code it to perform desired functions [33, 34]. This robotics system comes equipped with a suite of sensors, motors, a programmable brick, and myriad LEGO TECHNIC plastic components with an interlocking system of tubes and studs to build the mechanical structure and locomotion mechanism of a robot. LEGO's educational robotics product line additionally includes WeDo and SPIKE Essential systems for elementary school learners and SPIKE Prime system for middle school learners to cater to their varied needs and capabilities. Finally, VEX robots are similar to the LEGO robotics systems in that they are also endowed with a brick-based programming mechanism, can be programmed in a blocks- or text-based programming environment, and are available to serve the learning needs of various age groups [31]. While the structural components of the VEX robotics system for elementary and middle school age groups include plastic parts that snap together, for the high school age group they include aluminum parts that require mechanical assembly using nuts and bolts.

The modular kit-based robots commonly use pre-fabricated mechanical subsystems including chassis, wheels, legs, etc., that are typically assembled using fasteners. The assembly steps for such robotics systems are pre-determined, although they include options for customization based on the need of the learning activity. These robotics systems obviate the complexity and eliminate the design freedom entailed by the brick-based robots [31] concerning their mechanical structure and locomotion mechanism. Even as modular kit-based robotics systems shorten the mechanical assembly process, they often entail breadboarding to interface the various electronics, sensing, and actuation modules of a robot.

Thus, the modular kit-based robots offer opportunities to engage students in learning and practicing varied disciplinary content at a deeper level while also building and honing their prototyping skills. Some popular modular kit-based robots include Parallax's BoeBot [35], Elegoo's Car Robot [36], and Pololu's Romi [37], among others.

Pre-assembled robots can be used for various age groups and provide varied levels of coding options, for example, with blocks- and text-based programming using computers or tablets. For example, Microbric's Edison robot [38] comes equipped with LEDs, buttons, sensors, and motors. The Edison robot can be customized by using its studs and holes to interconnect with other Edison robots as well as with LEGO-compatible construction blocks [2, 38]. Yet another pre-assembled robot includes Scribbler from Parallax [35] that comes equipped with wheels, motors, sensors, etc., ready to operate with pre-programmed modes and capable of being programmed. Similarly, iRobot Create, based on the Roomba vacuum cleaner, is a robotics platform that comes ready to program and operate out of the box. KIWI robot [39] is yet another example of a pre-assembled educational robotics system. Finally, for young children who lack the kinesthetic capability to assemble a robotics creation or to use a computer for constructing a blocks-based robot program, there exist fully assembled physically coded educational robots that can enable learners to experience computational thinking and coding skills. To appeal to young learners, such robots are often endowed with an animated appearance and their functionality can be controlled with artifacts such as buttons, coded images (e.g., bar code, QR code), or physical blocks, that encapsulate a certain action to be performed by the robot. One example of such a robot is a BeeBot with an appealing exterior resembling a bee [1] that can be commanded using push buttons on top of the device. The previously mentioned KIWI robot [39] can be commanded by children with the use of physical code blocks. Dash [21] is yet another fully assembled robot that is appealing to children as it can be commanded using buttons, drawing paths, or creating programs. Finally, iCat [40] represents an embodied representation of a cat that can be enticing for children to interact with.

See Figures 2.1 (a)–(c) for images of some of the aforementioned robots. Moreover, see [2, 31] for additional examples of various brick-based, modular kit-based, and pre-assembled robotics systems. Applications of some of the robots in different learning settings are discussed in the next section.

3. Educational robotics: Roles and settings

The development and availability of various brick-based and modular kit-based robotics systems allow learners to design and assemble contemporary technological devices that they can control and with which they can learn, as opposed to passively using a ready-made technology device [41]. The act of assembling a

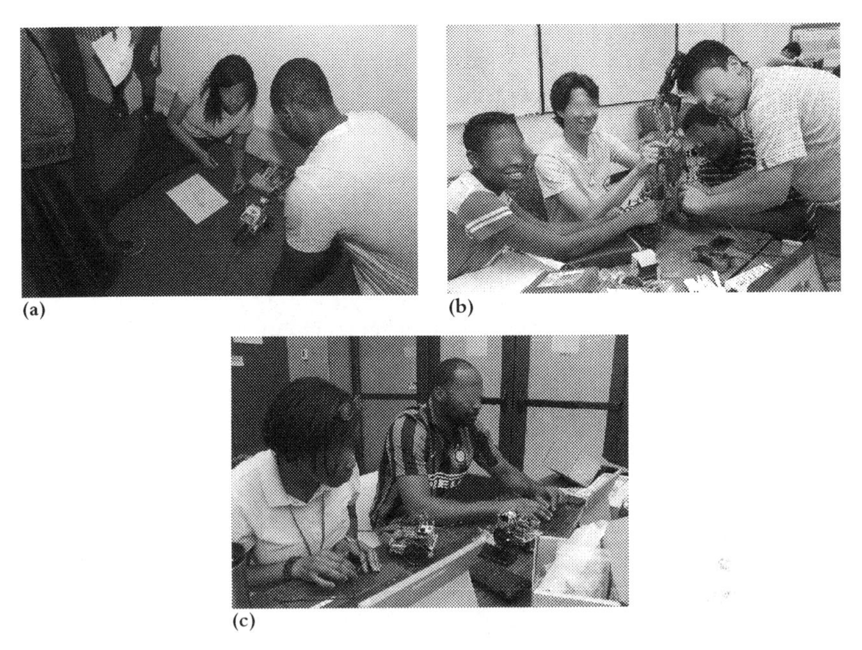

(a) (b) (c)

FIGURES 2.1 (a)–(c) Illustrative images of LEGO (a), VEX (b), and BoeBot (c) being used by learners.

robot and creating a program for it is grounded in the constructionist learning theory and it can guarantee the endurance and effectiveness of learning [21]. While the act of controlling the actions or behaviors of the robot and studying the robot-environment interaction are often not prioritized in educational settings, such activities are essential since they produce rich feedback that can promote student learning and understanding [41]. To summarize, the activities involved in constructing, programming, and controlling robots, embedded with high-quality instruction, can help engender creativity among learners, develop their problem-solving skills, endow them with a deeper understanding of concepts, etc., and thus such activities can play varied roles in the learning process of students. Based on a literature review on the educational applications of robots, this section presents two different ways to classify the usage of robots in educational settings [1, 10, 41]. The first set of classifications is based on the role of robots in the teaching and learning process and the second set of classifications is based on the role of robots in various educational contexts.

3.a. Role of robots in the learning process

Robots can play varied roles in the teaching and learning process based on the desired learning outcome, disciplinary content, types of learners, and educational settings. By performing the meta-analyses of research studies conducted in the last two decades, two comprehensive review articles [1, 10] have summarized various applications of robotics in education. Specifically, the three categories of robot usage based on their level of contribution in the learning process are identified as: a *tool* or learning object, a *peer* or learning companion, and a *tutor* or teaching assistant [1, 10].

The first category concerned with the use of robotics as a learning *tool* includes two primary ways of learning about robots [1, 10, 41]; first, using robotics as a discipline of its own and second, using robots to study other disciplines. In studying robotics as a stand-alone subject, learning tasks may include designing and constructing robot structure and locomotion mechanism; gaining knowledge about and purposefully utilizing sensors, actuators, electronics, and computing components; and creating and testing a program for robot operation; among others. Since robotics as a discipline and robot as a physical artifact are both inherently interdisciplinary, the process of building, instrumenting, and programming a robot entails implicitly learning, using, and reinforcing math, science, engineering, and computing concepts that can be applied to other disciplines. For example, under the PBL framework, utilizing the interdisciplinary nature of robotics can facilitate learning of STEM subjects. In its alternative and more explicit role, robotics can also be used as a learning tool that can support the learning of different disciplines at varied grade levels. Specifically, in this case, robotics-based explorations can allow students to gain alternative and concrete representations of abstract knowledge, for example, in science and math disciplines. When being used as a learning tool, to ensure its effectiveness in promoting learning, robotics must be meaningfully incorporated for instruction and exploration to align with the varied capabilities and diverse interests of learners [42].

According to [43], a cooperative learning setup has a variety of effects on and myriad benefits for student learning, including cognitive development, academic achievement, enhanced motivation, positive attitude, social competence, self-efficacy, and self-esteem, among others. Since cooperative learning is known to promote cognitive-academic development and socio-emotional growth among students, and it can engender an active learning atmosphere, it is valuable to simulate and embed such opportunities with the use of a robot as a learning companion or *peer*. Incorporation of robots as "social partners in learning" [44] can promote active social interactions leading to greater engagement and learning [44, 45], and developing a growth mindset [46]. Specifically, a study on preschoolers' language development using peer-like robots revealed that children's interactions with such robots can be leveraged for early language development [45]. The study of [45] employed a DragonBot robot [47] capable of facial expressions, sound, and speech, all using a smartphone device, and that is covered in green fur for aesthetics and to draw children's attention. Through eight sessions, the robot executed a storytelling game that introduced new vocabulary and allowed students to emulate language [45].

Finally, a robot can also be ascribed the role of a *tutor* wherein it functions and behaves like a teaching assistant, as in a teacher-student relationship, to produce supportive interactions instead of unidirectional information transfer [40]. Such a role for the robot can be facilitated through activities that include supporting social dialogue, adapting to serve students' learning needs, and helping in learning new material [1]. As an example, a language learning tutor robot application iCat was introduced in a primary school to provide varying levels of learning support [40]. The robot provided learning material based on students' learning needs and employed facial expressions to signify empathetic responses. The outcomes of the final assessment revealed that social supportiveness yields positive learning outcomes for language development and further indicated that the approach of drawing on social support may benefit educational media applications [40].

3.b. Robotics in educational settings

This classification of applications of robots relies on their use in informal *versus* formal educational settings [41]. In either of these contexts, robotics can be used as a tool, peer, or tutor, to maximize the learning opportunities for students. Informal learning environments, such as after-school academic clubs, represent flexible spaces and opportunities that allow students to have freedom and that additionally focus on evolving with the change in their learning needs [48]. After-school clubs, competitions, exhibitions, and summer camps are common avenues to learn about and through robotics in informal settings. Several national and international contests, such as the FIRST LEGO League, RoboCup, RoboParty, and RobotChallenge, attract participation from thousands of teams every year while seeking to catalyze innovation, teamwork, communication, and motivation to pursue STEM subjects among them [41].

To broaden student participation and enhance the impact of robotics activities, Rusk et al. [49] have suggested shifting focus from competitions to exhibitions. This strategy is deemed important as exhibitions foster collective participation, are more inclusive in nature, and allow greater room for expression and creativity than competitions. After-school programs and summer camps offer flexibility in curriculum selection, that is, it can range from a period of several weeks to an entire year, and render opportunities to connect students and teachers that are often missing in case of competitions [50]. Informal robotics-based learning programs have shown promise in enhancing student learning achievement [51], engagement, and communication [23], among others.

Alternatively, formal learning settings entail a structured environment, such as traditional school classrooms, where learning is organized to achieve intentional learning targets [48], often within specified durations of time, while being orchestrated by an instructor. Even as the informal learning environments are quite popular, they represent "free choice" settings that may have a stronger appeal to one gender than the others (e.g., competitions) [49] or that may represent barriers

(a)

(b)

FIGURES 2.2 (a)–(b) Illustrative images of robot use in informal (a) and formal (b) learning settings.

due to attendant financial, time, and transportation burden for some students (e.g., after-school, weekend, or summer programs, museums) [52], inhibiting the full potential that educational robotics has to offer. Moreover, the extracurricular nature of robotics contests has failed to make the use of robotics more central to K-12 science and math education. Introducing constructionist activities like educational robotics in formal school classrooms can increase access to learning and make the classrooms more equitable for students from all backgrounds [53]. To facilitate such an integration in formal classrooms, it is essential to align learning outcomes and skills acquired through robotics activities with the curriculum standards [53–55] and create professional development programs that support educators in successfully implementing such activities in their classrooms [9].

See Figures 2.2 (a)–(b) for some illustrative images of robot use in informal and formal learning environments.

4. Examples of applications of robots as a learning tool in teaching and learning

As seen in the previous sections, educational robotics can be incorporated into the teaching and learning process in a variety of ways based on the intended learning outcomes, student age groups and backgrounds, availability of resources, educational settings, etc. For instance, WaterBotics [56, 57] represents a curriculum and environment grounded in discovery-based learning that employs underwater robotics for engendering interest and engagement in STEM learning in both formal and informal settings. Using an iterative engineering design process, students prototype, test, and improve their solutions that can address real-life "missions," for example, rescuing a drowning person or cleaning a pollutant spill. After completing each mission, they showcase solutions to their peers. In another project, Teacher Education on Robotics-Enhanced Constructivist Pedagogical Methods (TERECoP) [58], a research team sought to create novel strategies and relevant learning environments that can prepare teachers to design robotics-based activities and implement them in their classrooms. Using the framework of social constructivism, the researchers created a collaborative community of practice for teachers to facilitate and sustain their professional preparation. In the following subsections, we provide three detailed examples of the use of robots as a learning tool in different learning settings.

4.a. Developmentally appropriate robotics kits and curriculum in early childhood classrooms [39]

This research study examined the student learning impact of an eight-week-long curriculum on robotics and programming in early education classrooms. By recognizing the increasing need for and interest in STEM education, the researchers focused their attention on the disciplines of technology and engineering, which often receive scant treatment in early childhood education, and sought to intentionally integrate the four disciplinary domains of STEM by investigating the promise of educational robotics in fostering cognitive, social, and fine motor skills. Conducted in the formal setting of classrooms, this research employed robotics as a learning tool, along with foundational programming concepts, to enable students to: develop and hone computational thinking skills, gain and practice ideas about designing everyday objects, and engender early exposure to STEM and programming to support them in overcoming gender stereotypes concerning STEM careers. In four classrooms from pre-kindergarten to second grade, 60 learners took part in the study by attending weekly hour-long lessons spanning over eight weeks. Research assistants serving on the research team taught the lessons while the classroom teachers supported them with curriculum implementation and learned themselves via observation for future enactment of the lessons.

This study utilized KIWI robots that seek to address the relative paucity of developmentally appropriate robotics systems for early childhood learners. The KIWI robotics system, developed by the DevTech Research Group at Tufts University, consists of both the robot hardware and a novel method for instructing its behavior. To render a robotics system for the developmentally relevant capabilities of early childhood learners, the researchers adhered to the following design considerations: physical and natural ease in connecting robot parts, ability to program the robot with minimal reliance on a computer, and means to endow the robot with varied artistic creations. To obviate the need for a computer and eliminate screen time, a KIWI robot is provided programming instructions using interconnecting physical wooden blocks affixed with circular barcodes. A barcode scanner mounted on the robot reads codes from each wooden block and interprets the corresponding instructions. This novel method for instructing the robot is referred to as a "Creative Hybrid Environment for Robot Programming (CHERP)." The robotics and programming curriculum was embedded in a comprehensive unit titled "Me and My Community" wherein each hands-on robotics lesson built on the learning from the preceding week ending in a capstone finale. In the first two weeks, the students began by learning about robot anatomy and the tangible block coding method, then attaching various robot parts and craft materials to a robot to represent themselves, and finally coding the robot to dance to the music of their choice. In the next four weeks, they learned about sound, light, and distance sensors and loops and conditionals concepts of programming while also practicing the same using their robots. The last two weeks were devoted to the students working in teams on a capstone project wherein they drew a map of their neighborhood, populated it with places of significance, and coded their robots to traverse the route and perform desired actions at chosen locations. Based on the student comfort level, the curriculum was paced appropriately. Whereas the pre-kindergarten learners required all the eight sessions to hone the foundational skills from the first two lessons, the first and second graders proceeded at a quicker pace and spent more time exploring sensors and complex coding concepts.

To assess student learning impact of this program concerning robotics and programming knowledge, two assessments were conducted individually with each student following the completion of the eight-week curriculum. The first assessment titled "Robots parts task" entailed a researcher individually asking each student to identify five components from the robot anatomy and recognize their functions, with each correct response receiving 1 point for a maximum possible score of 5. Next, under the second assessment titled "Solve-It," the researcher gave eight stories about a robot, tasked students to create CHERP-based programs to generate the corresponding behaviors for their robot, and graded their responses using a scoring rubric. The study reported positive outcomes for foundational robotics and programming knowledge among students, with students of higher grades performing better. Finally, the researchers concluded that in bringing robotics and programming to early childhood grades it is essential to employ developmentally appropriate robotics systems and to meaningfully integrate robotics and programming with other classroom-relevant curricular units.

4.b. An extra-curricular middle school engineering education program [59]

This study examined a two-year-long program, conducted in an informal learning setting, that utilized robotics as a learning tool for middle school students to: enhance their learning of STEM concepts, foster collaboration and workplace skills among them, afford them equitable STEM learning opportunities, and provide them knowledge to explore future STEM pathways. Recent decades have witnessed that the general public holds misperceptions of STEM disciplines and professionals, students lack enthusiasm for STEM studies and careers, and post-secondary STEM enrollments have been declining. In response, the study team sought to alleviate the situation by creating opportunities for students to: learn and practice STEM concepts while responding to complex problems in a socially responsible manner; cultivate and hone myriad professional skills by interacting with various stakeholders; acquire practical knowledge through cognitive apprenticeship to rout STEM stereotypes; and discover STEM pathways along with parents and guardians. Under the guidance of two teachers per school, approximately 100 middle schoolers from four schools participated in after-school extra-curricular activities for over 120 contact hours per year for two years. The educational programming included academic year activities for 78 contact hours and summer activities for 48 contact hours. To facilitate discovery-based, technology-focused learning, the program incorporated the LEGO Mindstorms NXT mobile robotics kit and PICO Cricket kits to fuse technology and art.

Grounded in an informal learning setting, the program rendered opportunities to innovate and pilot ideas and transfer learning to new contexts, all of which promote interest, motivation, and enthusiasm among students well beyond the outcomes of formal schooling. The program thoughtfully incorporated myriad evidence-based theoretical frameworks, such as PBL, engineering design process, 5E instructional model, and cognitive apprenticeship, to ensure that the students received a multi-disciplinary exposure and explored STEM concepts deeply. Specifically, the PBL framework was adopted to cultivate a collaborative and social environment imbued with authentic problems, framed in the real-world context of a STEM workplace, that spark critical thinking among students. Moreover, the engineering design process was incorporated to allow learners to design, test, and produce physical artifacts as solutions to design projects. Next, the 5E instructional model, with the five cyclical steps of *engage, explore, explain, elaborate*, and *evaluate*, was utilized to introduce students to discovery learning and to induce conceptual change. Finally, cognitive apprenticeship was embedded in the program to enable middle schoolers to obtain guidance in the authentic contexts of learning and workplace, as well as to gain familiarity with STEM pathways, from postsecondary students, faculty, professionals, and experts, all from STEM disciplines.

The program's multidisciplinary learning units included topics such as simulating the behaviors of desert tortoises using a LEGO Mindstorms robot model, studying the local cause and global effect by simulating large-scale chain reactions using a PICO Cricket kit, and examining the urban heat island by creating models to alleviate temperature effects. To elaborate on the project about simulation for a desert tortoise, the students began with gaining an understanding about tortoise behaviors in natural settings by: interacting with tortoises brought into the after-school program, visiting a zoo on a field trip, meeting with experts, and researching through internet resources. Having gained adequate understanding, students undertook the engineering design process, sketched a corresponding design, and followed it to construct their artifact using the LEGO Mindstorms robotics kit.

In addition to participating in the program's curriculum, students had opportunities to go on related field trips, parent engagement activities, and industry internships. Moreover, students participated in cognitive apprenticeships wherein they received mentorship from undergraduate engineering interns, industry professionals, and experts in the field. Pre-post assessments administered to measure STEM learning of participating students showed statistically significant differences in pre-post mean scores with higher mean scores on post-assessments, revealing enhanced learning in the content areas of each curriculum unit. A 'Draw an Engineer' test was also conducted at the start and end of the program wherein students were asked to draw an engineer based on their perceptions of engineering profession and respond to three related prompts. Analyses of the student drawings revealed a shift in their perceptions of engineers, while their initial drawings conveyed engineers as engaged in building mechanical apparatus, their final drawings envisioned an "engineer who thinks." Additionally, samples of student writing, focus group discussions with them, and observations helped in identifying student conceptions of and learning progress in STEM. The research team summarized the importance of their study by highlighting the need for continued organized efforts to design deep and long-term experiences that can educate students and families about future STEM pathways.

4.c. Workshops to expose learners to basics of robot programming in an informal learning setting [60, 61]

Realizing the need to develop age-appropriate robots, from early childhood to high school levels, without compromising with the rigor of learning, a research team from École Polytechnique Fédérale de Lausanne (EPFL) in collaboration with École Cantonal d'Arts de Lausanne (écal) designed a small and affordable Thymio II robot for use by children from different age groups. While the pre-assembled Thymio II robot is equipped with a repertoire of pre-programmed behaviors controlled using built-in buttons, it allows advanced programming options through a visual programming language (VPL) as well as a text-based scripting language. These robots were used to introduce learners to basic robot control activities as part of day-long workshops during the 2013 EPFL Robotics Festival. Specifically, under an informal learning setting, five different robotics workshops were conducted for learners of varied age groups and they were attended by approximately 500 participants.

The first workshop, *Discover Thymio II while playing*, targeted learners from four to eight years of age who did not have to perform any computer-based programming. Instead, the workshop participants used physical buttons embedded on the robot platform to trigger its pre-programmed operating modes, observed the corresponding behaviors to establish cause–effect relationships, and used their learning to solve assigned performance challenges for the robot. The second workshop, *Travel through space with Thymio II*, intended to serve learners from six to nine years of age who were knowledgeable about the pre-programmed behaviors of the robot. They were tasked with solving six space-related challenges using the cause–effect relationship for the robot's various buttons and its corresponding behavior. The third workshop, *Graphical programming with Thymio II*, catered to learners between eight and 11 years of age who were assumed to have prior exposure to computers. After interacting with the robot's pre-programmed modes and observing its responses, the students learned about the robot's sensors, actuators, and a visual programming method. Next, they were tasked to employ the VPL interface to solve challenges, such as commanding a robot to continue moving on a surface and to stop when it is lifted. The fourth workshop, *Introduction to programming with Thymio II*, involved participants from 12 to 15 years of age who were exposed to a text-based programming language to program the robot with variables and conditionals as well as for the use of sensors and actuators. Finally, the fifth and last workshop, *Advanced programming with Thymio II*, targeted participants from 14 to 18 years of age who were required to problem-solve for a challenge, which entailed motion, sensing, and decision-making, and code a solution for it independently.

Surveys were administered at the end of each workshop to understand its effect on the participants. An analysis of the survey data revealed that the participants attended the workshops primarily because they wanted 'to learn new things' and 'to have fun.' The participants generally appreciated the playful exploration activities with robots and, while the responses varied on the basis of age-based groups, a majority of them responded that they found the 'challenges stimulating' and 'learnt things that can be useful later.' In terms of perceived success, a majority of the participants agreed that they felt successful except for the fifth workshop attendees, possibly because it contained significantly more programming concepts, required independent work, and text-based programming is more complicated than VPL-based programming. Finally, for controlling the robot, more than 50% of participants from the first three workshops found it very easy to control the robot owing to the button-based or VPL-based programming. Participants from the fourth and fifth workshops found the control to be moderately easy. Overall, the research team considered that the workshops helped validate the design of the robot as useful, simple, and catering to a wide range of ages and genders.

5. Conclusion

The discipline of educational robotics is grounded in the fundamental learning theories of constructivism and constructionism [2]. While constructivist learning entails building on pre-existing conceptions to create new knowledge and understanding through mental processes, constructionist learning relies additionally on developing new knowledge through hands-on exploration with physical manipulatives, learning in a social environment, and situating learning in real contexts. In this vein, educational robotics endeavors to be learner-centered and it generates personally meaningful active learning experiences. Several pedagogical strategies have been examined to engender robotics-enhanced constructionist experiences, for example, project-, competition-, and inquiry-based learning [11]. The different types of robots used in disparate robotics-based learning activities can be classified on the basis of their physical design as: brick-based robots, modular kit-based robots, and pre-assembled robots. Moreover, these robots are additionally classified on the basis of their programming methods that include relatively simple and visual blocks-based programming, simple to advanced text-based programming, and physical blocks-based programming that obviates the reliance on computers and screens. Among the various brick-based robots, the LEGO Mindstorms robotics platform [33, 34] is popular in the K-12 educational environment. The ease of assembly of its mechanical components, plug-and-play capability of its sensors and actuators, and the ability of coding it using a blocks-based and a text-based programming environment have all contributed to the acceptance of LEGO Mindstorms among learners from upper elementary to high school grade levels. Alternatively, among the various kit-based robots, the Parallax's BoeBot [35] robotics platform is popular in secondary school environments, including in teacher professional development programs [62]. Finally, pre-assembled robots such as BeeBot [1], Edison [38], and Scribbler [35] offer varied programming methods, for example, all of them come equipped with pre-programmed behaviors while Edison and Scribbler can be programmed using blocks-based or text-based programming tools as well. Different types of robots can be used to create varied learning experiences based on their level of contribution in the learning process as a tool or learning object, a peer or learning companion, and a tutor or teaching assistant [1, 10] or, based on their use in informal and formal educational settings [41]. In recent years, various studies have examined the potential of educational robots for their ability to create varied learning experiences based on the type of robot, various uses of robots in achieving learning outcomes, different learning settings, etc. This chapter has summarized three recent studies to illustrate the applicability of robots as a learning tool in formal and informal settings. As seen throughout the chapter, understanding the role of robots and aligning it with purposeful learning goals can engender student-centered, collaborative, and active learning experiences.

6. Key takeaways

- Educational robotics is founded on significant learning theories of constructivism [7] and constructionism [8]. Both learning theories are learner-centered and focus on students creating their own learning experiences.
- Project-, competition-, and inquiry-based learning represent varied instructional approaches that can be applied to create constructionist learning environments for robotics-based learning [11].

- Numerous types of robotics platforms can be used for educational activities. While there can be different ways to classify the types of robots, two key ways to classify them include based on their physical design and programming method [1, 2, 31].
- Robots can be used on the basis of their level of contribution in the learning process as a tool or learning object, a peer or learning companion, and a tutor or teaching assistant [1, 10]. They can also be classified on the basis of their use in informal and formal educational settings [41].

References

1. Mubin, O., Stevens, C.J., Shahid, S., Al Mahmud, A., and Dong, J.-J., A review of the applicability of robots in education. *Journal of Technology in Education and Learning*, 2013.1: p. 1–7.
2. Eguchi, A., Theories and practices behind educational robotics for all, in *Handbook of Research on Using Educational Robotics to Facilitate Student Learning*. 2021, Hershey, PA: IGI Global. p. 68–106.
3. Eguchi, A., Robotics as a learning tool for educational transformation, in *Proceedings of International Workshop Teaching Robotics, Teaching with Robotics and International Conference Robotics in Education*. 2014, Padova, Italy. p. 27–34; Available from: https://www.terecop.eu/TRTWR-RIE2014/files/00_WFr1/00_WFr1_04.pdf.
4. Petre, M. and Price, B., Using robotics to motivate 'back door' learning. *Education and Information Technologies*, 2004.9(2): p. 147–158.
5. Ruiz-del-Solar, J. and Avilés, R., Robotics courses for children as a motivation tool: The Chilean experience. *IEEE Transactions on Education*, 2004.47(4): p. 474–480.
6. Williams, K., Igel, I., Poveda, R., Kapila, V., and Iskander, M., Enriching K-12 science and mathematics education using LEGOs. *Advances in Engineering Education*, 2012.3(2): p. 1–27.
7. Wadsworth, B.J., *Piaget's Theory of Cognitive and Affective Development: Foundations of Constructivism*. 1996, White Plains, NY: Longman Publishing.
8. Papert, S., *Mindstorms: Children, Computers, and Powerful Ideas*. 1980, New York, NY: Basic Books, Inc.
9. You, H.S., Chacko, S.M., and Kapila, V., Examining the effectiveness of a professional development program: Integration of educational robotics into science and mathematics curricula. *Journal of Science Education and Technology*, 2021.30(4): p. 567–581.
10. Anwar, S., Bascou, N.A., Menekse, M., and Kardgar, A., A systematic review of studies on educational robotics. *Journal of Pre-College Engineering Education Research (J-PEER)*, 2019.9(2): p. 19–42.
11. Altin, H. and Pedaste, M., Learning approaches to applying robotics in science education. *Journal of Baltic Science Education*, 2013.12(3): p. 365–377.
12. Ackermann, E., Piaget's constructivism, Papert's constructionism: What's the difference? in *Proceedings of Constructivism: Uses and Perspectives in Education*. 2001, Geneva, Switzerland: Research Center in Education. p. 85–94; Available from: https://learning.media.mit.edu/content/publications/EA.Piaget%20_%20Papert.pdf.
13. Schreiber, L.M. and Valle, B.E., Social constructivist teaching strategies in the small group classroom. *Small Group Research*, 2013.44(4): p. 395–411.

14. Morado, M.F., Melo, A.E., and Jarman, A., Learning by making: A framework to revisit practices in a constructionist learning environment. *British Journal of Educational Technology*, 2021.52(3): p. 1093–1115.
15. Kafai, Y.B., Constructionism, in *The Cambridge Handbook of the Learning Sciences*, R.K. Sawyer, Editor. 2005, Cambridge, UK: Cambridge University Press. p. 35–46.
16. Stager, G., Papertian constructionism and the design of productive contexts for learning, in *Proceedings of EuroLogo*. 2005, Warsaw, Poland. p. 43–53.
17. Mikropoulos, T.A. and Bellou, I., Educational robotics as mindtools. *Themes in Science and Technology Education*, 2013.6(1): p. 5–14.
18. You, H.S., Chacko, S.M., and Kapila, V., Teaching science with technology: Scientific and engineering practices of middle school science teachers engaged in a robot-integrated professional development program (Fundamental), in *Proceedings of ASEE Annual Conference and Exposition*. 2019, Tampa, FL; Available from: https://peer.asee.org/33353.
19. PBLWorks Buck Institute for Education, *What is PBL?* Available from: www.pblworks.org/what-is-pbl.
20. Blumenfeld, P.C., Soloway, E., Marx, R.W., Krajcik, J.S., Guzdial, M., and Palincsar, A., Motivating project-based learning: Sustaining the doing, supporting the learning. *Educational Psychologist*, 1991.26(3–4): p. 369–398.
21. Eguchi, A., Educational robotics theories and practice: Tips for how to do it right, in *Robots in K-12 Education: A New Technology for Learning*, B.S. Barker, Editor. 2012, Hershey, PA: IGI Global.
22. Issa, G., Hussain, S.M., and Al-Bahadili, H., Competition-based learning: A model for the integration of competitions with project-based learning using open source LMS. *International Journal of Information and Communication Technology Education (IJICTE)*, 2014.10(1): p. 1–13.
23. Chris, C., Learning with FIRST LEGO league, in *Proceedings of Society for Information Technology and Teacher Education (SITE) International Conference*. 2013, New Orleans, LA. p. 5118–5124.
24. National Research Council, *National Science Education Standards*. 1996, Washington, DC: National Academies Press.
25. Dodge, M.M., *The Effect of the 5E Instructional Model on Student Engagement and Transfer of Knowledge in a 9th Grade Environmental Science Differentiated Classroom*, Master of Science—Science Education, Montana State University July 2017, (Thesis).
26. National Research Council, *Inquiry and the National Science Education Standards: A Guide for Teaching and Learning*. 2000, Washington, DC: National Academy Press.
27. Blancas, M., Valero, C., Vouloutsi, V., Mura, A., and Verschure, P.F., Educational robotics: A journey, not a destination, in *Handbook of Research on Using Educational Robotics to Facilitate Student Learning*, S. Papadakis and M. Kalogiannakis, Editors. 2021, Hershey, PA: IGI Global. p. 41–67.
28. NGSS, *Next Generation Science Standards (NGSS): For States, By States*. 2013, Washington, DC: The National Academies Press; Available from: www.nextgenscience.org/.
29. CoCFNSS, *A Framework for K-12 Science Education: Practices, Crosscutting Concepts, and Core Ideas*. Committee on Conceptual Framework for the New K-12 Science Education Standards (CoCFNSS). 2012, Washington, DC: National Academies Press.
30. Dewey, J., *Dewey on Education*. 1959, New York, NY: Teachers College Press.
31. Karim, M.E., Lemaignan, S., and Mondada, F., A review: Can robots reshape K-12 STEM education? in *Proceedings of IEEE International Workshop on Advanced Robotics and its Social Impacts (ARSO)*. 2015, Lyon, France. p. 1–8.

32. Resnick, M., Martin, F., Sargent, R., and Silverman, B., Programmable bricks: Toys to think with. *IBM Systems Journal*, 1996.35(3.4): p. 443–452.
33. Martin, F.G., *Robotics Explorations: A Hands-On Introduction to Engineering*. 2001, Upper Saddle River, NJ: Prentice Hall.
34. Perdue, D.J., *The Unofficial LEGO Mindstorms NXT Inventor's Guide*. 2007, San Francisco, CA: No Starch Press.
35. Parallax Inc., *All Robotics*. 2022; Available from: www.parallax.com/product-category/all-robotics/.
36. Elegoo, *Robot Kits*. 2022; Available from: www.elegoo.com/collections/robot-kits.
37. Pololu Robotics & Electronics, *Robot Kits*. 2022; Available from: www.pololu.com/category/2/robot-kits.
38. Microbric, *Edison Robot*. 2022; Available from: https://meetedison.com/.
39. Sullivan, A. and Bers, M.U., Robotics in the early childhood classroom: Learning outcomes from an 8-week robotics curriculum in pre-kindergarten through second grade. *International Journal of Technology and Design Education*, 2016.26(1): p. 3–20.
40. Saerbeck, M., Schut, T., Bartneck, C., and Janse, M.D., Expressive robots in education: Varying the degree of social supportive behavior of a robotic tutor, in *Proceedings of* ACM *SIGCHI Conference on Human Factors in Computing Systems*. 2010, Atlanta, GA. p. 1613–1622.
41. Alimisis, D. and Kynigos, C., Constructionism and robotics in education, in *Teacher Education on Robotic-enhanced Constructivist Pedagogical Methods*, D. Alimisis, Editor. 2009, Greece: School of Pedagogical and Technological Education (ASPETE). p. 11–26.
42. Alimisis, D., Educational robotics: Open questions and new challenges. *Themes in Science and Technology Education*, 2013.6(1): p. 63–71.
43. Nastasi, B.K. and Clements, D.H., Research on cooperative learning: Implications for practice. *School Psychology Review*, 1991.20(1): p. 110–131.
44. Okita, S.Y., Ng-Thow-Hing, V., and Sarvadevabhatla, R., Learning together: ASIMO developing an interactive learning partnership with children, in *Proceedings of IEEE International Symposium on Robot and Human Interactive Communication (RO-MAN)*. 2009, Toyama, Japan. p. 1125–1130.
45. Kory-Westlund, J.M. and Breazeal, C., A long-term study of young children's rapport, social emulation, and language learning with a peer-like robot playmate in preschool. *Frontiers in Robotics and AI*, 2019.6: p. 81: 1–17.
46. Park, H.W., Rosenberg-Kima, R., Rosenberg, M., Gordon, G., and Breazeal, C., Growing growth mindset with a social robot peer, in *Proceedings of ACM/IEEE International Conference on Human-Robot Interaction (HRI)*. 2017, Vienna, Austria. p. 137–145.
47. Kory, J.M., Jeong, S., and Breazeal, C.L., Robotic learning companions for early language development, in *Proceedings of ACM on International Conference on Multimodal Interaction*. 2013, Sydney, Australia. p. 71–72.
48. Callanan, M., Cervantes, C., and Loomis, M., Informal learning. *Wiley Interdisciplinary Reviews: Cognitive Science*, 2011.2(6): p. 646–655.
49. Rusk, N., Resnick, M., Berg, R., and Pezalla-Granlund, M., New pathways into robotics: Strategies for broadening participation. *Journal of Science Education and Technology*, 2008.17(1): p. 59–69.
50. Karp, T. and Maloney, P., Exciting young students in grades K-8 about STEM through an afterschool robotics challenge. *American Journal of Engineering Education*, 2013.4(1): p. 39–54.

51. Barker, B.S. and Ansorge, J., Robotics as means to increase achievement scores in an informal learning environment. *Journal of Research on Technology in Education*, 2007.39(3): p. 229–243.
52. Migus, L.H., *Broadening Access to STEM Learning Through Out-of-School Learning Environments*. 2014, Washington, DC: National Research Council Committee on Successful Out-of-School STEM Learning. p. 1–17; Available from: https://sites.nationalacademies.org/cs/groups/dbassesite/documents/webpage/dbasse_089995.pdf.
53. Eguchi, A., Bringing robotics in classrooms, in *Robotics in STEM Education*, M.S. Khine, Editor. 2017, Cham, Switzerland: Springer. p. 3–31.
54. Ghosh, S., Krishnan, V.J., Rajguru, S.B., and Kapila, V., Middle school teacher professional development in creating a NGSS-plus-5E robotics curriculum (Fundamental), in *Proceedings of ASEE Annual Conference and Exposition*. 2019, Tampa, FL; Available from: https://peer.asee.org/33108.
55. You, H.S., Chacko, S.M., Rajguru, S.B., and Kapila, V., Designing robotics-based science lessons aligned with the three dimensions of NGSS-plus-5E model: A content analysis (Fundamental), in *Proceedings of ASEE Annual Conference and Exposition*. 2019, Tampa, FL; Available from: https://peer.asee.org/32622.
56. McKay, M.M., Lowes, S., Tirthali, D., and Camins, A.H., Student learning of STEM concepts using a challenge-based robotics curriculum, in *Proceedings of ASEE Annual Conference and Exposition*. 2015, Seattle, WA; Available from: https://peer.asee.org/24756.
57. Holahan, P., McKay, M., Sayres, J., Lowes, S., Camins, A., and McGrath, B., *WaterBotics: A Novel Engineering Design Curriculum for Formal and Informal Educational Settings*. 2015; Available from: https://waterbotics.org/media/uploads/doc/waterbotics_monograph.pdf.
58. Alimisis, D., Moro, M., Arlegui, J., Pina, A., Frangou, S., and Papanikolaou, K., Robotics & constructivism in education: The TERECoP project, in *Proceedings of EuroLogo*. 2007, Bratislava, Slovakia. p. 19–24.
59. Ganesh, T., Thieken, J., Baker, D., Krause, S., Roberts, C., Elser, M., Taylor, W., Golden, J., Middleton, J., and Kurpius, S.R., Learning through engineering design and practice: Implementation and impact of a middle school engineering education program, in *Proceedings of ASEE Annual Conference and Exposition*. 2010, Louisville, KY; Available from: https://strategy.asee.org/16972.
60. Riedo, F., Chevalier, M., Magnenat, S., and Mondada, F., Thymio II, a robot that grows wiser with children, in *Proceedings of IEEE Workshop on Advanced Robotics and its Social Impacts (ARSO)*. 2013, Tokyo, Japan. p. 187–193.
61. Riedo, F., Rétornaz, P., Bergeron, L., Nyffeler, N., and Mondada, F., A two years informal learning experience using the Thymio robot, in *Advances in Autonomous Mini Robots*, U. Rückert, J. Sitte, and F. Werner, Editors. 2012, Berlin, Germany: Springer. p. 37–48.
62. Kapila, V. and Lee, S.-H., Science and mechatronics-aided research for teachers. *IEEE Control Systems Magazine*, 2004.24(5): p. 24–30.

3

TEACHING STEM WITH ROBOTICS

Synopsis of a research-guided program

1. Introduction

Educators, policymakers, and educational institutions unanimously agree on the vital importance of integrating 21st-century skills into school curriculum to equip learners for success in the rapidly evolving landscape of global innovation economy and the emerging future of work [1]. This clear and growing attention to the preparation and training of the workforce for the future has led to a rapidly accelerating infusion of technologies such as adaptive learning, artificial intelligence, augmented reality, virtual reality, etc., in classrooms [2]. Nonetheless, students find it challenging to stay engaged and motivated in learning when they deem technological interventions to be devoid of personally meaningful contexts and irrelevant to the learning tasks. A systematic approach for engaging students in science, technology, engineering, and mathematics (STEM) disciplines and enhancing their interest in STEM career pathways requires the creation of active explorations, founded on novel curricula and instructional methods, that embed authentic and meaningful learning experiences.

Using educational technologies, such as robotics, to embed meaningful and motivating engineering and technology contexts in science and math courses constitutes a compelling strategy as it can engender engaging learning experiences, foster conceptual understanding, and support the application of learning to personally relevant scenarios. The two critical challenges in integrating these technologies in classrooms include (1) lack of curricula that embed contemporary technologies to foster authentic STEM learning and (2) inadequate teacher preparation in using technologies effectively for STEM education. To surmount the aforementioned twin challenges for successfully integrating educational robotics in science and math learning, a recent research effort performed

DOI: 10.4324/b23177-4

design-based research (DBR) to develop novel robotics-based STEM curriculum and a collaborative teacher professional development model. Using iterative design–implement–refine cycles, the professional development program cultivated hands-on experiences and collaborative activities that encouraged teachers to co-design a robotics-infused STEM curriculum. The project utilized robust theoretical frameworks including technological, pedagogical, and content knowledge (TPACK), 5E instructional model, and project-based learning (PBL), among others to prepare teachers and encourage them in using robotics in a pedagogically effective manner.

This chapter gives a synopsis of the multi-year effort that sought to address the lack of curricula and inadequate teacher preparation to facilitate effective adoption of educational robotics in STEM classrooms. First, the chapter explains the need to design authentic STEM experiences and the rationale behind the use of educational robotics in producing effective STEM learning. Next, the chapter shares the details of the research effort including descriptions of its theoretical underpinnings, design of the program, examples from implementation, and its outcomes. The research project produced positive results in learning, engagement, and motivation among both teachers and students. Throughout all activities described below, engineering and education researchers purposefully partnered with middle school teachers to collectively design, enact, and refine the project and determine its outcomes, while being guided by the relevant theoretical constructs.

2. Need for authentic STEM learning experiences

Exponential growth of information technology, prevalence of low-cost computing hardware, improved functionality of engineered products, and a global expansion of communication and transportation networks are key characteristics of our modern world. Today's global innovation economy affects not only the products and industries in the engineering and technology sectors, but is also integral to myriad everyday affairs ranging from healthcare to finance and entertainment to education. The constant reshaping of our daily lives by scientific discoveries and technological innovations calls for students to acquire scientific and engineering savvy and become well-versed with STEM disciplines, so that they can grow with and respond to the pace of this accelerating advancement. To fulfill the demands for a tech-savvy future workforce, which can ensure the United States (US) to maintain its competitive edge in the innovation economy, requires our educational institutions to employ novel curricula and instructional methodologies for STEM education. Even as the ever-advancing technological landscape creates an urgent demand for the STEM workforce, only 20% of the country's high school graduates are prepared to pursue STEM majors in college [3], racially diverse students are less likely to pursue careers in STEM, and students from low income and gender-diverse backgrounds have limited access to informal STEM programs [4]. The aforementioned trends reveal that our educational system is unable to satisfy the prevailing demands and, thus, requires interventions that can support the students and teachers in acquiring skills that advance STEM learning.

Being at the forefront of educating, preparing, and influencing student achievement, teachers themselves must possess the requisite knowledge and skills to equip students in STEM disciplines [5]. Even when teachers are cognizant of the importance of engendering authentic STEM learning experiences, they often find it challenging to embed contemporary technologies in their context, for example, subject matter, classroom, and student interest, and to stay abreast with the evolving hardware and software features of various technological tools [6]. Their lack of familiarity with the technology tools acts as a barrier for teachers in building engaging and relevant STEM learning classrooms [7]. Consequently, exposure to uninteresting curriculum and outmoded instructional methods make it challenging for students to appreciate the interdisciplinary nature of STEM and instead gives them siloed experiences into the discipline. These dynamics prevent students from having any meaningful opportunities to improve their STEM knowledge or skills with relevance to and applications in the modern workplace [8].

Representation of knowledge in an abstract manner [7], without being linked to personal experiences of students, causes them to fail in gaining a deep understanding of STEM topics and feel disengaged from STEM learning. According to research [9], 30% of students lose interest in science disciplines by fourth grade and approximately 50% of students become disinterested in pursuing STEM careers by eighth grade. Despite being digitally adept, today's students lack the understanding of foundational science and engineering concepts [10]. Knowing the lack of interest and perceived value of STEM among students [11], it is necessary to conceive, design, and enact hands-on experiential learning activities that are deemed personally meaningful by them and that generate interest, enthusiasm, and engagement. Doing so necessitates a curriculum that incorporates familiar and contemporary technologies [12, 13], is relevant [14], and infuses a purposeful blend of the four disciplines of STEM. Moreover, alleviating students' deficits in STEM understanding and interest additionally requires professional development programs that can empower teachers to develop comfort with contemporary technology tools and meaningfully incorporate them in STEM instruction. The novel curriculum and pedagogical techniques can build on students' technological interest and imagination while purposefully developing their STEM knowledge [10].

3. Rationale for robotics in STEM education

Integration of educational robotics in middle school STEM learning is a compelling proposition since it has the potential to: make abstract concepts accessible through easy visualizations, stimulate hands-on learning [15], and foster engagement and motivation in learning [10]. Learning through robots cultivates a setting that encourages students to engage in constructing knowledge through concrete manipulatives as well as understanding abstract concepts tangibly and reflecting on their real-world implications. As delineated in Chapter 2, this process and type of learning stems from the learning theories of Jean Piaget's constructivism and Seymour Papert's constructionism [10, 16]. Specifically, performing constructionist activities with robots provides students with exploration opportunities that can allow them to conceptualize, build, and evaluate their ideas [17]. Next Generation Science Standards (NGSS) [18] offer a unified framework that incorporates three dimensions of science learning, including engineering design and practices, and has connections with Common Core State Standards for Mathematics (CCSSM) [19]. Additionally, the use of robotics as an instructional tool provides hands-on experiential learning opportunities that promote collaboration, critical thinking, and problem-solving [20]. Thus, employing NGSS and CCSSM along with educational robotics as a learning tool is a promising strategy to develop novel STEM experiences that afford active, contextual, and engaging learning opportunities for all students.

While many popular robotics competitions provide means for students to build varied life and professional skills (e.g., communication, collaboration, and critical thinking, among others), the extracurricular nature of these contests has prevented robotics from becoming an integral component of K-12 STEM education [10]. Moreover, various enrichment-based K-12 robotics programs (e.g., after-school clubs, summer camps) require the presence of support staff, which can limit their potential for scaling and sustainability [10]. Prior attempts to address the aforementioned constraints and promote broad-scale adoption of K-12 robotics programs have included teacher professional development interventions that seek to positively contribute to the development of robotics curricula and teacher training [21]. For example, the Teacher Education on Robotics-Enhanced Constructivist Pedagogical Methods (TERECop) project aimed to provide teachers with opportunities for professional development to design and implement computer-based robotics activities [22]. Yet another professional development intervention, Robotics and Engineering Education-Fostering the Conceptual Understanding of Science (RE^2-Fo-CUS), provided teachers with engineering design-based modules to educate them about human behaviors affecting animals as well as to build their domain-based content knowledge and self-efficacy [23]. Lastly, another professional development workshop employed the Kids Invent with Imagination (KIWI) robot kits [24] to enhance the knowledge of teachers for robotics, engineering, programming, and pedagogy for the early childhood education environment.

4. Overview, theoretical background, and project design

This section provides an overview of a multi-year educational research project that was implemented in the classrooms of 22 middle schools across an urban school

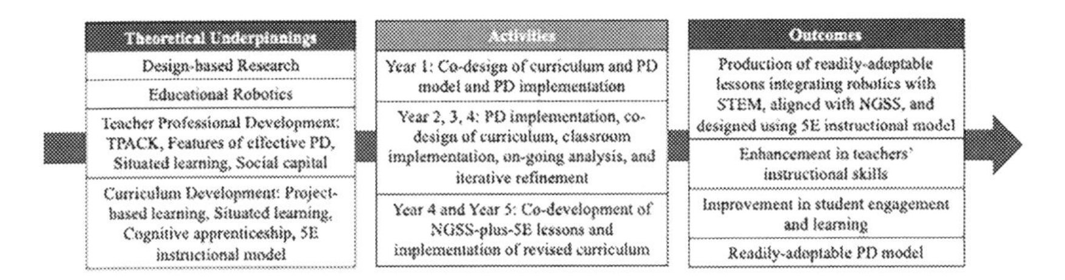

FIGURE 3.1 Elements of the research project.

district in the northeastern US, impacting over 2,000 students and 44 teachers. The project was conceptualized to address two major challenges: lack of curricula to ensure the adoption of robotics in science and math classrooms and positively influence student learning, and lack of teacher preparation and training for constructing effective models of knowledge that help build capacity to incorporate robotics in the classroom. Beneficiaries of the project included middle school science and math teachers as well as their students. The teachers participated in summer professional development programs, co-developed robotics-based lessons, contributed to varied activities focused on improving the curriculum and instructional design, and implemented the newly developed lessons to engage students in robotics-based experiential learning. The project sought to utilize DBR [25] to iteratively develop and refine a curriculum model that employs robotics as a pedagogical tool to support learning for middle school science and math disciplines. Furthermore, the project sought to design a research-guided professional development model to foster skills and attitudes for incorporating robotics-based learning among middle school science and math teachers. See Figure 3.1 that encapsulates the main aspects of the research project.

4.a. Theoretical underpinnings

The design, development, implementation, and analysis of the project's robotics-based STEM curriculum and teacher professional development were based on several purposely selected theoretical constructs that rendered opportunities to advance research and exploration in this field. The design of the project was primarily guided by: (1) the DBR paradigm [25] to co-design and co-create the curriculum and professional development in partnership with middle school teachers; (2) the TPACK framework [26] to develop and deploy authentic and effective robotics-based lessons and professional development; and (3) features of effective professional development to provide teachers active, coherent, and collaborative learning experiences [27, 28]. In addition to utilizing the aforementioned constructs in the iterative development of curriculum and enhanced learning of teachers, the project progressively researched and incorporated additional evidence-based approaches, such as PBL to impart hands-on experiences [29]; situated learning [30, 31] to contextualize the overall learning experience;

cognitive apprenticeship to learn from role models [32]; 5E instructional model [33] to develop NGSS-plus-5E lesson plans focused on conceptual change; and social capital model [34, 35] to create collaborative learning opportunities and lesson plans. In this subsection, we provide brief overviews of the various research constructs and practical techniques and highlight their applications in the context of this project. Moreover, the following chapters delve into these various frameworks and methods in greater detail with myriad examples, including the ones focused on robotics-based STEM learning experiences.

(i) Design-based research (DBR)

DBR is a methodology that targets the improvement of educational practices and evolution of educational principles through an iterative cycle of analysis, design, implementation, and evaluation performed by researchers and practitioners in a collaborative partnership [36]. The use of DBR can advance learning theories through empirical evidence obtained via cycles of iterative refinements. Specifically, DBR can contribute to the understanding of how and why the designed learning and curriculum approaches work while concurrently offering ideas for redesign. Moreover, DBR can help in establishing how individual outcomes are related to specific educational solutions, while also validating their effectiveness. Thus, DBR can aid in studying the efficacy of educational products, processes, programs, and policies [37] under multiple domains, for example, curriculum development, media and technology in education, instructional design, and teacher professional development [38, 39]. The integration of DBR in educational studies is based on an understanding that teachers play a central role in curriculum-resource integration; teaching is a design-based activity wherein curricular materials afford and constrain teacher action; and teachers use materials and resources differently [40].

In this project, DBR was employed for curriculum design and professional development by including teachers as design partners in a participatory design endeavor. For the curriculum design effort, DBR can enhance the authenticity and efficacy of curriculum implementation through the deliberate inclusion of teacher perspectives in its adaptation and enactment. This can generate field-tested and iteratively improved artifacts and practices for classroom viability and design principles that contribute to the effective development of curricular materials. The use of DBR can also allow examination and discovery of limitations in the prevailing curriculum to support student learning of disciplinary core ideas (DCI), crosscutting concepts (CCC), and science and engineering practices (SEP) that are suggested in NGSS and CCSSM. For the teacher professional development effort, DBR can deepen teachers' contextual learning of science and math and allow them to explore engineering design. Furthermore, it can permit teachers to foster motivating and engaging environments that empower students to take ownership of their learning. Finally, DBR can allow examination of features of curricula, practices, and instruction that support teacher learning and facilitate the integration of robotics in science and math teaching. See Chapter 4 for examples that illustrate applications of the DBR method for creating robotics-based curriculum and professional development programming.

(ii) Technological, pedagogical, and content knowledge (TPACK)

To meaningfully infuse technology in disciplinary pedagogy, teachers must be endowed with opportunities that develop their TPACK. Being equipped with TPACK enables teachers to design learning experiences that render alternative and accessible representations of abstract knowledge for learners [26], instead of using technology in pedagogically misaligned and unsophisticated ways. The construct of TPACK facilitates uncovering of the synergistic interactions between three knowledge domains, namely, technology, pedagogy, and content, and it allows technology to be *intentionally* used as an instructional tool for creating and facilitating novel representations of knowledge [26]. Whereas the prior TPACK research explored digital technologies (videos, websites, online courses, graphing calculators, etc.), this effort considered a novel instantiation of TPACK to robotics. Specifically, the TPACK framework informed the project's robotics-based curricula and professional development activities. Teachers were mentored to discover the use of robotics in science and math context to develop novel and varied representations of disciplinary content. Teachers used a variety of pedagogical approaches such as inquiry-based or project-based learning and incorporated their knowledge of students' prior understandings and backgrounds in the design, development, and testing of robotics-based, hands-on activities to address science and math content and build curriculum. We further adopted criteria suggested in literature for designing effective technology-integrated lessons [41, 42]. See Chapter 6 for examples that illustrate applications of the TPACK framework in robotics-focused professional development and in developing robotics-based learning units.

(iii) Features of effective professional development

One of the challenges in implementing effective technology-based lessons is teachers' lack of experience with advanced technology tools and their pedagogical integration [6]. To render an environment that is conducive to teacher learning about technology-infused lessons, features of effective professional development must be embedded in the design of the program. Specifically, effective professional development programs: focus on content, incorporate active learning opportunities, infuse coherence, sustain over sufficient durations of time, and foster collective participation [27, 28]. Under this project, teachers were provided opportunities for hands-on exploration of various aspects of robotics, such as construction, motion, actuation, sensing, and programming, and relevant tools such as sensors, actuators, programmable brick, etc. [43]. Teachers co-created lessons and experiential learning activities that employ these technological tools effectively and meaningfully. We identified the need to support teachers in building knowledge of robotics fundamentals in a context where they concurrently integrate science and math concepts and construct their TPACK; get comfortable with and apply robotics to teach STEM; and have a high likelihood of adopting robotics in their classrooms. Additionally, in the context of teacher professional development, helping individuals cultivate social capital can yield advantageous returns [44]. Specifically, the quality of personal relationships [45] can facilitate greater access to specialized knowledge [35]. Moreover, investing in the social network can contribute to cross-functional team effectiveness [46] (e.g., teams of science and math teachers [35]). Thus, this project employed the social capital model [34, 35] to surface a variety of thoughts and perspectives that can be employed to collaboratively develop and deploy student-centered lessons that integrate robotics with STEM education (see a detailed example in Chapter 5, Section 6).

(iv) Project-based learning (PBL)

PBL represents a teaching and learning approach, contextualized in a real-world project, that endows students with opportunities to gain knowledge and skills through active exploration, examination, and analysis of a personally relevant challenge [29]. This method supports long-term retention and deeper understanding of concepts, improved collaboration and problem-solving skills, and enhanced learning attitudes [47]. When integrated with educational robotics, PBL can support learners in synthesizing knowledge from multiple domains [48], fostering higher-order cognitive skills [49], and examining and addressing a design challenge through robot design and programming. Moreover, integrating robotics to provide hands-on experiential learning permits just-in-time feedback, which can help learners to link the robot behavior to their design choices as well as enhance their design, and create an environment to promote self-directed learning [47]. Building on these strengths, the project's instructional methodology leveraged the PBL framework to create real-world hands-on learning scenarios through robotics. For instance, in a project-based math class, a teacher introduced the concept of parabolas by tasking students to construct a ping-pong ball launcher that can maximize the height of the launched ball [50] (see Figure 3.2). This book consists of several additional examples and details pertaining to the integration of PBL with robotics in STEM education, see, for example, Subsection 4.b of Chapter 1 and Section 4 of Chapter 9, among others.

FIGURE 3.2 Middle schoolers engaged in project-based learning with their ping pong launcher robot.

(v) Situated learning

Situated learning transpires when students are engaged in problem-solving through real-world situations, collaborative group work, and reflection [30, 31]. In the context of enhancing teacher expertise to embed technology in STEM pedagogy, situated learning can contribute to developing teachers' instructional practices and analyzing their beliefs about teaching with robots. Thus, for the project's professional development program, we initiated situated learning by having teachers partake in the curriculum first as learners themselves, deliberately performing robotics-based experiential learning activities to view the curriculum from a student perspective, and then reflecting on it and revising it. Having gained this knowledge and experience, the teachers next had opportunities to create and test robotics-based science and math lessons embedded with specific situations, which were deemed to have potential to enhance student learning and engagement. See Subsection 4.b of Chapter 4 and Section 5 of Chapter 5 for examples on the use of situated learning for teacher professional development and classroom instruction.

(vi) Cognitive apprenticeship

Cognitive apprenticeship refers to an instructional theory that seeks to bridge the gap between formal education and real-life work situations by supporting learners to develop cognitive skills required for professional practices [32]. Enactment of this method in a classroom requires the teacher to embed socially and functionally contextualized real-world scenarios of the profession in the curriculum [51]. Moreover, it entails allowing students to gain knowledge and skills in authentic contexts through social interactions under the guidance of domain experts [6]. In the professional development program of this project, cognitive apprenticeship was purposefully embedded by engaging teachers to collaboratively design, test, and iteratively refine robotics-based learning activities under the mentorship and support of robotics and education researchers (see details in Chapter 5, Section 5). Moreover, several classroom lessons were delivered by the teachers in the presence of robotics researchers who guided the learners through robotics activities under the apprenticeship model (see details in Chapter 4, Section 4).

(vii) 5E instructional model

The 5E instructional model represents a constructivist pedagogical approach that seeks to bring about a conceptual change in the learner [33] and develop their higher-order thinking skills [52]. Through the five phases of *engage, explore, explain, elaborate,* and *evaluate*, learners become aware of their prior conceptions, explore new ways of thinking about a problem, go deeper in their understanding of concepts, grow adept at applying their knowledge to different situations, and become capable of presenting their learning [33]. The 5E model can be applied to design curriculum materials, curriculum frameworks, assessment guidelines, teacher professional development programs, and informal education experiences [53]. The project employed the 5E method (see Chapter 9) to develop STEM lessons that integrated robotics and national standards such as the three dimensions of NGSS and CCSSM. Through summer professional development workshops, we introduced the framework to teachers and supported them in developing their own lessons that integrated the NGSS and 5E model and were eventually termed *NGSS-plus-5E* robotics-based lessons.

4.b. Project design and activities

The project sought to create a state-of-the-art and contextual educational robotics curriculum as well as a program for training teachers to build and implement such a curriculum. To achieve the intended outcomes and to study them for broad dissemination to the larger education community, the project designed and conducted extensive research. The research agenda included two broad goals. First, the research aimed to identify the fundamental characteristics of robotics-based curriculum to effectively teach and learn middle school science and math. Second, the research targeted to ascertain and implement the essential elements of professional development to build teachers' TPACK for using educational robotics. In support of these overarching goals of research, the project sought to understand how teachers integrate their professional development experience in STEM classrooms and what are the necessary student prerequisites to employ robotics in classrooms.

Following is a summary of the plans and activities that were pursued for the project's six-year implementation cycle. During the beginning phase of the project, the curriculum and professional development model was co-created by researchers and teachers. After a thorough revision process involving pilot testing and consultation with teachers under the DBR framework, the professional development program was implemented pursuant to which the curriculum was enacted in the classrooms. The on-going research and analysis triggered further enhancements that led to a revised NGSS-plus-5E incorporated curriculum. A summary of the project with timelines is provided in Table A.1 of Appendix A.

Design of curriculum and professional development material: In the first year of the project, five science and five math lessons were developed to alleviate content-specific pedagogical challenges by purposefully incorporating robotics-based explorations [31]. The preliminary design of lessons addressed the five domains of the robotics learning sequence, that is, construction, motion, actuation, sensing, and programming [43]. The robotics learning sequence provided rich opportunities to embed active learning of myriad science and math concepts and it served as a link to connect different subjects [54]. The 10 lessons were intentionally designed to treat the aforementioned five domains. During a three-week professional development pilot in summer, a design-based approach was utilized wherein a cohort of four teachers first experienced the lesson and its activities as students, then they engaged in reflection and discussion to analyze the preliminary curriculum design for its alignment with standards and classroom contexts, and finally, they expanded on it based on their perspectives. In this manner, the teachers iteratively revised the lessons for a tighter alignment with the school science and math curriculum. Based on their knowledge of middle school science and math curricula and students' backgrounds, as well as their newly gained knowledge of robotics, they proposed developing new lessons [31] as well as eliminating and replacing several lessons deemed to have ineffective technology integration with alternative lessons that had the potential to offer novel representations of concepts [43]. The resulting outcomes gave the project team a new perspective on the opportunities and challenges of employing robotics as an educational tool and revisions needed to improve the professional development program design. See Table B.1 in Appendix B for a list of various robotics-based science and math lessons, aligned with the Common Core State Standards, developed over the project duration.

Classroom implementation and revised professional development workshop: Having participated in the pilot professional development program in summer, during the following academic year, the four teachers implemented the curriculum in their classrooms and suggested revisions based on their experiences. Based on their feedback [31], the professional development workshop was revised and conducted next summer for a second cohort of 20 teachers. The teachers were involved in situated learning through hands-on activities that helped them gain confidence in integrating robotics in science and math teaching. Consistent with TPACK, they were encouraged to reflect on pedagogical elements central to integrating robotics and active learning. They were given multiple opportunities to revise the curriculum before employing it for classroom use. Next, during the academic year, they implemented the newly developed robotics lessons in their classrooms while being supported by the project personnel through two to three school visits and follow-up meetings every month. The school visits by the project staff and teacher visits with the project team enabled to: identify the scaffolds needed by teachers for lesson implementation [55], obtain feedback on curriculum and pedagogy [31], evaluate teacher practices [56], and revise the curriculum for the following year of implementation [31]. This entire sequence of professional development workshop and classroom implementation was repeated for a third cohort of 23 teachers.

Curriculum revision and NGSS-plus-5E professional development workshop: In the fourth year of the project, a revised curriculum aligned with the three dimensions of NGSS under the 5E instructional model [35] was developed and the subsequent lessons were termed NGSS-plus-5E lessons. A three-week summer professional development program was conducted with revised curriculum for six teachers who had previously participated in the professional development workshops. These teachers in turn served as master teachers to deliver the curriculum to 11 new teachers. The master teachers integrated the NGSS-plus-5E robotics curriculum in their classrooms to identify additional refinement needs and learn about improvements in the overall project model. This data was analyzed to examine if teachers can use robotics effectively to integrate disciplinary content knowledge, infuse NGSS and CCSSM in the curriculum, and support student learning with contextual and creative learning activities [57, 58]. Based on the analysis and further refinements, another NGSS-plus-5E professional development workshop was conducted next summer with three master teachers who mentored 20 new teachers in exploring and iterating the new lessons. See Table B.2 in Appendix B for a list of NGSS-plus-5E robotics-based lessons developed under this project.

5. Illustrative examples from implementation

Aligning the design of project's professional development program with the features of effective professional development [27, 28]**:** Through numerous real-world examples, the project's professional development workshops encouraged teachers to look beyond one correct answer and be comfortable with a level of engineering analysis [59] to create a richer and more nuanced learning environment for students. Considering the *content focus* of the project, the facilitators used their prior experience in delivering content-focused (mechatronics) professional development workshops [60–62] and concentrated on modeling research-informed and content-specific (robotics) instructional practices. To prepare teachers to effectively employ robotics as a tool to teach middle school science and math, the professional development program addressed both the content knowledge of robotics fundamentals as well as the methods for applying robotics to teach science and math. During the three-weeklong summer professional development workshops, teachers experienced hands-on robotics-based learning through *active participation* that simultaneously engendered project-based learning. The preliminary curriculum design produced lessons that coincided with the learning sequence of robotics while treating the middle school science and math content [43, 54]. Under this approach, as participants learned middle school science and math concepts, they concurrently received exposure to important robotics concepts, including construction, motion, actuation, sensing, and programming.

Next, to produce a *coherent alignment* with the NGSS and CCSSM, the curriculum design purposefully focused on creating NGSS-plus-5E robotics-based lessons [35, 57]. Teachers' awareness about what their students may find challenging has implications for how they may integrate robotics-based engineering in their classes and spur students with higher learning expectations. Thus, in consultation with teachers, we intentionally selected a variety of topics (e.g., antibiotic resistance, energy, and friction, among others) that are deemed pedagogically challenging, within the context of the traditional classroom instructional methods, and that can be taught by utilizing the affordances of robotics as a suitable learning tool that provides means of alternate representation and engagement [63]. The teachers were encouraged to focus on understanding and using robotics concepts most pertinent with teaching science and math, instead of having to learn robotics concepts that may be too advanced for their academic background and may have limited use in the classroom. The three-week *duration* of summer workshop, academic year follow-up, cohort model involving a science and a math teacher from the same school, and *collective participation*, among others, were intentional attributes of project design inspired by prior research on features of effective professional development.

Using educational robotics to teach an abstract science topic [63]: Often students do not comprehend potential and kinetic energy from a system's perspective, that is, a system can simultaneously have these two forms of energy. In the ideal case, the system energy is conserved, and thus these forms of energy transform from one to another during motion. Using robotics provides an opportunity wherein such a system can be represented in an appealing and visual manner. Specifically, a zipliner robot [63, 64] can be created that moves under the influence of gravity along the inclined zipline on a path that can be changed by altering the zipline's initial and final heights. Using a robot equipped with an ultrasonic sensor and programmed to acquire sensor data using wireless communication, its height and speed can be autonomously logged and plotted as it traverses the zipline (see Figure 3.3). With the acquired data, the potential and kinetic energies

FIGURE 3.3 A zipline robot to illustrate the concepts of conservation of energy and transformation of forms of energy.

of the robot can be calculated at any two or more locations to visualize and verify that the total mechanical energy is conserved. Consistent with Earle's and Ferdig's criteria for effective technology integration [41, 42], a lesson plan was developed for such a zipline robot. The students were first introduced to the definitions of potential and kinetic energies from a systems perspective using demonstrations and discussions. Using visual illustrations and representations of a ball falling from a height, the students were exposed to the idea of transformation of potential energy to kinetic energy. Applying the new knowledge to a scenario, the students generated hypotheses about the energy states of a person traveling down on an inclined zipline. The students used a zipliner robot mounted with an ultrasonic sensor to gather data on kinetic and potential energies as it moved down the inclined path under the force of gravity. By formulating their hypotheses and performing experimental investigations, the students gained familiarity with the concept of energy conversion. In this manner, a disciplinary concept, which is ill-suited for manual representation and is not well understood by students, is given an easily accessible representation with a robot. The zipline robot lesson is one of more than 30 robotics-based STEM lessons that were originally created under another project. See Table B.3 in Appendix B for a list of these additional lessons. Chapters 8 and 9 also showcase STEM learning through several robotics-based lessons. See Figures 3.4 (a)–(b) illustrating middle school students engaged in a variety of robotics-based learning activities.

(a)

(b)

FIGURES 3.4 (a)–(b) Middle schoolers engaging in collaborative learning through robotics activities: gear ratio and speed (a) and cell cycle (b).

6. Project outcomes and recommendations

A qualitative and quantitative analysis of the project was conducted with the data collected using in-person and video observations of instruction enactment in class; logs of monthly debriefing sessions; and concept inventories, pre-post-tests, and validated assessments such as Test of Science Related Attitudes (TOSRA) [65], TPACK self-efficacy test [66], NGSS Evaluating the Quality of Instructional Products (EQuIP) [67], and 5E Inquiry Lesson Plan (ILP) [68] rubrics. Utilization of DBR throughout the implementation led to the evolution of principles that can be employed in similar settings. In alignment with the project's emphases on creating, testing, and refining curriculum and professional development, the following outcomes and recommendations were identified, as these can be beneficial for future adopters of robotics in STEM teaching and learning.

Characteristics of robotics-based curricula to effectively teach and learn science and math: Robotics-based learning was successful in generating and retaining student attention in the presence of a relevant and engaging problem. This was showcased by an improvement from 35% average pretest score to 72% average post-test score based on a disciplinary content knowledge test [54]. The utilization of PBL engendered opportunities for students to conceptualize solutions concretely and the integration of cognitive apprenticeship aided them in enhancing their solutions [54]. The implementation of lessons recommended some practices to best refine and contextualize the curriculum including, integration of elements based on students' backgrounds, knowledge, and skills; activating students' prior knowledge; splitting activities into small manageable components; having whole-class instruction and discussion prior to small group work; etc. [54]. The extent of alignment between the NGSS and robotics-based lessons was validated using EQuIP and 5E ILP rubrics for two lessons and the results revealed 'adequate alignment' for one lesson and suggestions to improve the second lesson. Thus, it was deemed feasible to align robotics lessons with the 5E model and three dimensions of NGSS [57].

Professional development model that builds teachers' TPACK and connects robotics with science and math: The professional development program focused on providing teachers with opportunities to actively model and practice technical skills for classroom enactment of lessons and exploring meaningful integration of robotics with science and math learning content [31, 43]. See Figures 3.5 (a)–(c) that illustrate educators engaged in collaborative learning and discussions. The analysis of pre- and post-tests of teachers' disciplinary content knowledge demonstrated a 24.9% increase with an average of 46.1% mastery prior to and 57.6% mastery after participating in the professional development [21]. Moreover, on the TPACK self-efficacy test [66], it was noticed that two factors, that is, confidence and outcome expectancy, had statistically

(a)

(b)

(c)

FIGURES 3.5 (a)–(c) Educators engaging in lessons as learners and brainstorming collaboratively: mini golf (a), genetic mutation (b), and educational research concepts (c).

significant increase from pre- to post-professional development [21]. To build teachers' TPACK for effective use, it has been recommended that a professional development program focus on enhancing the disciplinary content knowledge of teachers while aiding them to gain an understanding of relevant technology (in this case, robots) and its practical applications in education [43]. To further their TPACK, teachers may engage in learning communities, analyze their lessons for effective technology integration, and get familiar with varied technology tools for learning [69]. In employing the 5E framework, teachers must be well aware of students' prior conceptions and use activities to surface them prior to engaging in lesson tasks to successfully implement robotics-based learning [35]. In alignment with situated learning and PBL [31], teachers should prepare relevant real-world scenarios and embed them in lesson plans to generate hands-on engagement with and concrete understanding of abstract concepts [56].

Prerequisites and perceptions to effectively use robotics in science and math learning: Computational thinking ability is an essential requirement for success in robotics-based learning activities and can be built up by engaging students in tasks that empower them to use thinking and reasoning skills, creativity and imagination, problem-solving, block-based programming, systems thinking, etc. Behavioral and social aptitudes, laboratory and technical skills, engineering and design skills, and disciplinary content knowledge were also identified as essential for learners to collaborate in their social networks, manage resources, troubleshoot challenges, and build on their previous knowledge. To develop these prerequisites, students can be trained on the engineering concepts of robots and social and behavioral attitudes [55]. A careful examination of lessons for robotics-based explorations as well as a thorough preparation of and familiarity with hardware and software elements is essential to create a positive perception for educational robotics among students. In this vein, concerning student engagement, a shift from negative and neutral toward positive affective states was noticed amongst 7% of the participants. The most common positive perceptions were 'enjoyment,' 'support for conceptual understanding,' 'hands-on,' etc. As evidenced from [70], classroom-related factors such as pedagogical strategies and management of robotics-based educational content were mediating factors that contributed to the formation of student perceptions. See Chapter 7 for additional details.

7. Conclusion

This chapter presents an overview of the theoretical underpinnings, design, and outcomes of a multi-year project performed by a team of researchers in collaboration with teachers. Developed under the DBR paradigm and employing features of effective professional development and TPACK, the project was successful in enhancing teachers' knowledge and skills to co-design a curriculum and incorporate it in their STEM classrooms. Hands-on and situated curriculum was designed and deployed in classrooms, and positive student outcomes in terms of engagement, affect, and disciplinary content knowledge were noted. The project was also critical in creating teacher communities for learning and creating a robust teacher professional development model. Upon research and retrospective analysis, a few recommendations on teacher practices and successful instructional methodologies were identified and published in scholarly articles. Taken together, the project outcomes led to the development of curriculum, professional development model, and instructional strategies that can be adopted to design effective, relevant, and meaningful robotics-infused STEM learning for students.

8. Key takeaways

- The researchers theorized a robotics-based STEM education effort to address the lack of curricula, which can facilitate meaningful integration of robotics in science and math learning, and teacher training to promote the adoption of robotics for classroom instruction.

- The multi-year project was built using DBR to co-design a curriculum and a professional development model with teachers and by infusing features of effective professional development in the design, all to intentionally develop their TPACK. DBR is a research method that aims to enhance educational practices via iterative cycles of analysis, design, development, and implementation in collaboration with researchers and practitioners.
- Characteristics of robotics-based curricula to effectively teach and learn science and math include the presence of a relevant and engaging problem, project-based learning, ensuring that the curriculum is contextual and aligned with standards, breaking the lessons into small components, and facilitating whole group discussions prior to small group work, among others.
- Professional development model elements that build teachers' TPACK and connect robotics with science and math include: sharing and promoting knowledge of disciplinary content and technology, hands-on learning, co-design of situated curriculum, creating and sustaining a learning community, collaborative peer learning, and engaging teachers to experience the curriculum as learners.
- Necessary prerequisites to effectively use robotics in science and math learning include students' prior exposure to computational thinking, building of behavioral and social aptitude, laboratory and technical skills, engineering and design skills, and disciplinary content knowledge. Moreover, exposure to robotics may shift students' perceptions from negative/neutral to positive.

References

1. National Science Foundation, *Preparing the Next Generation of STEM Innovators: Identifying and Developing Our Nation's Human Capital*. 2010, Arlington, VA: National Science Foundation.
2. Google for Education, *Future of The Classroom: Emerging Trends in K-12 Global Education*; Available from: https://services.google.com/fh/files/misc/future_of_the_classroom_emerging_trends_in_k12_education.pdf.
3. Herman, A., America's STEM crisis threatens our national security. *American Affairs*, 2019.3(1): p. 127–148.
4. Grossman, J.M. and Porche, M.V., Perceived gender and racial/ethnic barriers to STEM success. *Urban Education*, 2014.49(6): p. 698–727.
5. Sanders, W.L. and Rivers, J.C., *Cumulative and Residual Effects of Teachers on Future Student Academic Achievement*. 1996, Knoxville, TN: University of Tennessee.
6. You, H.S. and Kapila, V., Effectiveness of professional development: Integration of educational robotics into science and math curricula, in *Proceedings of ASEE Annual Conference and Exposition*. 2017, Columbus, OH; Available from: https://peer.asee.org/28207.

7. AeA, *Losing the Competitive Advantage? The Challenge for Science and Technology in the United States*. 2005, Washington, DC: American Electronics Association.
8. Sobhan, S., Yakubov, N., Kapila, V., Iskander, M., and Kriftcher, N., Modern sensing and computerized data acquisition technology in high school physics labs, in *Advances in Computer, Information, and Systems Sciences, and Engineering*, K. Elleithy, T. Sobh, A. Mahmood, M. Iskander, M. Karim, Editors. 2007, Dordrecht: Springer. p. 441–448.
9. Hurst, M.A., Polinsky, N., Haden, C.A., Levine, S.C., and Uttal, D.H., Leveraging research on informal learning to inform policy on promoting early STEM. *Social Policy Report*, 2019.32(3): p. 1–33.
10. Williams, K., Igel, I., Poveda, R., Kapila, V., and Iskander, M., Enriching K-12 science and mathematics education using LEGOs. *Advances in Engineering Education*, 2012.3(2): p. 1–27.
11. Weiner, B., Why the U.S. has a STEM shortage and how we fix it (part 1). *Recruiting Daily*, November 6, 2018; Available from: https://recruitingdaily.com/why-the-u-s-has-a-stem-shortage-and-how-we-fix-it-part-1/.
12. Benitti, F.B.V., Exploring the educational potential of robotics in schools: A systematic review. *Computers & Education*, 2012.58(3): p. 978–988.
13. Riojas, M., Lysecky, S., and Rozenblit, J., Educational technologies for precollege engineering education. *IEEE Transactions on Learning Technologies*, 2011.5(1): p. 20–37.
14. Young, J.L., Young, J.R., and Ford, D.Y., Culturally relevant STEM out-of-school time: A rationale to support gifted girls of color. *Roeper Review*, 2019.41(1): p. 8–19.
15. Eguchi, A., Robotics as a learning tool for educational transformation, in *Proceedings of International Workshop Teaching Robotics, Teaching with Robotics and International Conference Robotics in Education*. 2014, Padova, Italy. p. 27–34; Available from: https://www.terecop.eu/TRTWR-RIE2014/files/00_WFr1/00_WFr1_04.pdf.
16. Ackermann, E., Piaget's constructivism, Papert's constructionism: What's the difference? in *Proceedings of Constructivism: Uses and Perspectives in Education*. 2001, Geneva, Switzerland: Research Center in Education. p. 85–94; Available from: https://learning.media.mit.edu/content/publications/EA.Piaget%20_%20Papert.pdf.
17. Alimisis, D., Educational robotics: Open questions and new challenges. *Themes in Science and Technology Education*, 2013.6(1): p. 63–71.
18. CoCFNSS, *A Framework for K-12 Science Education: Practices, Crosscutting Concepts, and Core Ideas*. Committee on Conceptual Framework for the New K-12 Science Education Standards (CoCFNSS). 2012, Washington, DC: National Academies Press.
19. Common Core State Standards Initiative, *Common Core State Standards for Mathematics*. 2010; Available from: https://learning.ccsso.org/wp-content/uploads/2022/11/Math_Standards1.pdf.
20. Blancas, M., Valero, C., Vouloutsi, V., Mura, A., and Verschure, P.F., Educational robotics: A journey, not a destination, in *Handbook of Research on Using Educational Robotics to Facilitate Student Learning*, S. Papadakis and M. Kalogiannakis, Editors. 2021, Hershey, PA: IGI Global. p. 41–67.
21. You, H.S., Chacko, S.M., and Kapila, V., Examining the effectiveness of a professional development program: Integration of educational robotics into science and mathematics curricula. *Journal of Science Education and Technology*, 2021: p. 1–15.
22. Alimisis, D., Moro, M., Arlegui, J., Pina, A., Frangou, S., and Papanikolaou, K., Robotics & constructivism in education: The TERECoP project, in *Proceedings of EuroLogo*. 2007, Bratislava, Slovakia. p. 19–24.

23. Schnittka, C., Turner, G.E., Colvin, R.W., and Ewald, M.L. A state-wide professional development program in engineering with science and math teachers in Alabama: Fostering conceptual understandings of STEM, in *Proceedings of ASEE Annual Conference and Exposition*. 2014, Indianapolis, IN; Available from: https://peer.asee.org/19998.
24. Bers, M., Seddighin, S., and Sullivan, A., Ready for robotics: Bringing together the T and E of STEM in early childhood teacher education. *Journal of Technology and Teacher Education*, 2013.21(3): p. 355–377.
25. The Design-Based Research Collective, Design-based research: An emerging paradigm for educational inquiry. *Educational Researcher*, 2003.32(1): p. 5–8.
26. Mishra, P. and Koehler, M.J., Technological pedagogical content knowledge: A framework for teacher knowledge. *Teachers College Record*, 2006.108(6): p. 1017–1054.
27. Birman, B.F., Desimone, L., Porter, A.C., and Garet, M.S., Designing professional development that works. *Educational Leadership*, 2000.57(8): p. 28–33.
28. Garet, M.S., Porter, A.C., Desimone, L., Birman, B.F., and Yoon, K.S., What makes professional development effective? Results from a national sample of teachers. *American Educational Research Journal*, 2001.38(4): p. 915–945.
29. PBLWorks Buck Institute for Education, *What is PBL?* Available from: www.pblworks.org/what-is-pbl.
30. Anderson, J.R., Reder, L.M., and Simon, H.A., Situated learning and education. *Educational Researcher*, 1996.25(4): p. 5–11.
31. Moorhead, M., Elliott, C.H., Listman, J.B., Milne, C.E., and Kapila, V., Professional development through situated learning techniques adapted with design-based research, in *Proceedings of ASEE Annual Conference and Exposition*. 2016, New Orleans, LA; Available from: https://peer.asee.org/25967.
32. Collins, A., Cognitive apprenticeship and instructional technology, in *Educational Values and Cognitive Instruction: Implications for Reform*, L. Idol and B.F. Jones, Editors. 1991, Mahwah, NJ: Lawrence Erlbaum Associates, Inc. p. 121–138.
33. Tanner, K.D., Order matters: Using the 5E model to align teaching with how people learn. *CBE Life Sciences Education*, 2010.9(3): p. 159–164.
34. Bourdieu, P., The forms of capital, in *Handbook of Theory and Research for the Sociology of Education*, J. Richardson, Editor. 1986, New York, NY: Greenwood Publishing Group. p. 241–258.
35. Ghosh, S., Krishnan, V.J., Rajguru, S.B., and Kapila, V., Middle school teacher professional development in creating a NGSS-plus-5E robotics curriculum (Fundamental), in *Proceedings of ASEE Annual Conference and Exposition*. 2019, Tampa, FL; Available from: https://peer.asee.org/33108.
36. Wang, F. and Hannafin, M.J., Design-based research and technology-enhanced learning environments. *Educational Technology Research and Development*, 2005.53(4): p. 5–23.
37. McKenney, S. and Reeves, T.C., About educational design research, in *Conducting Educational Design Research*. 2018, New York, NY: Routledge. p. 7–30.
38. van den Akker, J., Principles and methods of development research, in *Design Approaches and Tools in Education and Training*, J. van den Akker, et al., Editors. 1999, Dordrecht: Springer Netherlands. p. 1–14.
39. McKenney, S. and Reeves, T.C., Contributions to theory and practice: Concepts and examples, in *Conducting Educational Design Research*. 2018, New York, NY: Routledge. p. 30–60.

40. Ertmer, P.A., Addressing first-and second-order barriers to change: Strategies for technology integration. *Educational Technology Research and Development*, 1999.47(4): p. 47–61.
41. Earle, R., The integration of instructional technology into public education: Promises and challenges. *Education Technology Magazine*, 2002.42(1): p. 5–13.
42. Ferdig, R.E., Assessing technologies for teaching and learning: Understanding the importance of technological pedagogical content knowledge. *British Journal of Educational Technology*, 2006.37(5): p. 749–760.
43. Brill, A.S., Elliot, C.H., Listman, J.B., Milne, C.E., and Kapila, V., Middle school teachers' evolution of TPACK understanding through professional development, in *Proceedings of ASEE Annual Conference and Exposition*. 2016, New Orleans, LA; Available from: https://peer.asee.org/25720.
44. Lin, N., Building a network theory of social capital, in *Social Capital: Theory and Research*, N. Lin, K. Cook, and R.S. Burt, Editors. 2017, New York, NY: Routledge. p. 3–28.
45. Bybee, R.W., NGSS and the next generation of science teachers. *Journal of Science Teacher Education*, 2014.25(2): p. 211–221.
46. Adler, P.S. and Kwon, S.-W., Social capital: Prospects for a new concept. *Academy of Management Review*, 2002.27(1): p. 17–40.
47. Eguchi, A., Educational robotics theories and practice: Tips for how to do it right, in *Robots in K-12 Education: A New Technology for Learning*, B.S. Barker, Editor. 2012, Hershey, PA: IGI Global. p. 1–30.
48. Dym, C.L., Agogino, A.M., Eris, O., Frey, D.D., and Leifer, L.J., Engineering design thinking, teaching, and learning. *Journal of Engineering Education*, 2005.94(1): p. 103–120.
49. Muldoon, J., Phamduy, P.T., Le Grand, R., Kapila, V., and Iskander, M.G., Connecting cognitive domains of Bloom's taxonomy and robotics to promote learning in K-12 environment, in *Proceedings of ASEE Annual Conference and Exposition*. 2013, Atlanta, GA; Available from: https://peer.asee.org/19343.
50. Ghosh, S., Sabouri, P., and Kapila, V., Examining the role of LEGO robots as artifacts in STEM classrooms, in *Proceedings of ASEE Annual Conference and Exposition*. 2020, Virtual; Available from: https://peer.asee.org/34620.
51. Collins, A., Brown, J.S., and Newman, S.E., Cognitive apprenticeship: Teaching the craft of reading, writing and mathematics. *Thinking: The Journal of Philosophy for Children*, 1988.8(1): p. 2–10.
52. Boddy, N., Watson, K., and Aubusson, P., A trial of the five Es: A referent model for constructivist teaching and learning. *Research in Science Education*, 2003.33(1): p. 27–42.
53. Bybee, R.W., Taylor, J.A., Gardner, A., Scotter, P.V., Powell, J.C., Westbrook, A., and Landes, N., *The BSCS 5E Instructional Model: Origins and Effectiveness*. 2016, Colorado Springs, CO: Office of Science Education, National Institutes of Health.
54. Moorhead, M., Listman, J.B., and Kapila, V. A robotics-focused instructional framework for design-based research in middle school classrooms, in *Proceedings of ASEE Annual Conference and Exposition*. 2015, Seattle, WA; Available from: https://peer.asee.org/23444.
55. Rahman, S., M.M., Chacko, S.M., Rajguru, S.B., and Kapila, V., Fundamental: Determining prerequisites for middle school students to participate in robotics-based STEM lessons: A computational thinking approach, in *Proceedings of ASEE Annual*

Conference and Exposition. 2018, Salt Lake City, UT; Available from: https://peer.asee.org/30549.

56. Krishnan, V.J., Rajguru, S.B., and Kapila, V., Analyzing successful teaching practices in middle school science and math classrooms when using robotics (Fundamental), in *Proceedings of ASEE Annual Conference and Exposition*. 2019, Tampa, FL; Available from: https://peer.asee.org/32092.
57. You, H.S., Chacko, S.M., Rajguru, S.B., and Kapila, V., Designing robotics-based science lessons aligned with the three dimensions of NGSS-plus-5E model: A content analysis (Fundamental), in *Proceedings of ASEE Annual Conference and Exposition*. 2019, Tampa, FL; Available from: https://peer.asee.org/32622.
58. You, H.S., Chacko, S.M., and Kapila, V., Teaching science with technology: Scientific and engineering practices of middle school science teachers engaged in a robot-integrated professional development program (Fundamental), in *Proceedings of ASEE Annual Conference and Exposition*. 2019, Tampa, FL; Available from: https://peer.asee.org/33353.
59. Brophy, S., Klein, S., Portsmore, M., and Rogers, C., Advancing engineering education in P-12 classrooms. *Journal of Engineering Education*, 2008.97(3): p. 369–387.
60. Kapila, V. and Lee, S.-H., Science and mechatronics-aided research for teachers. *IEEE Control Systems Magazine*, 2004.24(5): p. 24–30.
61. Kapila, V., Research experience for teachers site: A professional development project for teachers, in *Proceedings of ASEE Annual Conference and Exposition*. 2010, Louisville, KY; Available from: https://peer.asee.org/16321.
62. Krishnamoorthy, S.P., Rajguru, S.B., and Kapila, V., Fundamental: A teacher professional development program in engineering research with entrepreneurship and industry experiences, in *Proceedings of ASEE Annual Conference and Exposition*. 2018, Salt Lake City, UT; Available from: https://peer.asee.org/30547.
63. Brill, A.S., Listman, J.B., and Kapila, V., Using robotics as the technological foundation for the TPACK framework in K-12 classrooms, in *Proceedings of ASEE Annual Conference and Exposition*. 2015, Seattle, WA; Available from: https://peer.asee.org/25015.
64. Yuvienco, C., *A Zipliner's Delight. A Classroom Lesson from AMPS/CBRI*. 2011; Available from: http://engineering.nyu.edu/gk12/amps-cbri/pdf/classroom%20activities/Carlo/a_zipliners_delight.pdf.
65. Fraser, B.J., Development of a test of science-related attitudes. *Science Education*, 1978.62(4): p. 509–515.
66. Schmidt, D.A., Baran, E., Thompson, A.D., Mishra, P., Koehler, M.J., and Shin, T.S., Technological pedagogical content knowledge (TPACK) the development and validation of an assessment instrument for preservice teachers. *Journal of Research on Technology in Education*, 2009.42(2): p. 123–149.
67. Achieve and National Science Teachers Association, *EQuIP Rubric for Lessons and Units: Science*. 2014; Available from: www.nextgenscience.org/resources/equip-rubric-science.
68. Goldston, M.J., Dantzler, J., Day, J., and Webb, B., A psychometric approach to the development of a 5E lesson plan scoring instrument for inquiry-based teaching. *Journal of Science Teacher Education*, 2013.24(3): p. 527–551.
69. Rahman, S.M.M., Krishnan, V.J., and Kapila, V., Exploring the dynamic nature of TPACK framework in teaching STEM using robotics in middle school classrooms, in

Proceedings of ASEE Annual Conference and Exposition. 2017, Columbus, OH; Available from: https://peer.asee.org/28336.

70. Ghosh, S., Rajguru, S.B., and Kapila, V., Investigating classroom-related factors that influence student perceptions of LEGO robots as educational tools in middle schools (Fundamental) in *Proceedings of ASEE Annual Conference and Exposition*. 2019, Tampa, FL; Available from: https://peer.asee.org/33023.

PART II

Theory, design, and implementation

With a goal to deepen the readers' understanding of designing robotics-enhanced learning experiences for teachers and students, this part of the book details three theoretical constructs through three chapters. The fourth chapter of the book, *Design-based Research for Robotics-enhanced Learning Environments*, showcases the underlying concepts, applications, and examples of usage in developing robotics-focused teacher professional development programs and curricula through design-based research. Next, the fifth chapter, *Effective Professional Development for Robotics-focused Learning Environments*, provides a detailed background, examples from the literature, and applications of the core features that make a technology-enhanced professional development program effective. Finally, the sixth chapter of the book, *Applying TPACK to Design for Robotics-enhanced Learning*, details the framework of technological, pedagogical, and content knowledge, a concept involving the required set of seven knowledge domains that help in building purposeful technology-enhanced learning experiences. Each of the three concepts in this part provides detailed examples, case studies, outcomes, and recommendations for educators to partake in similar explorations.

DOI: 10.4324/b23177-5

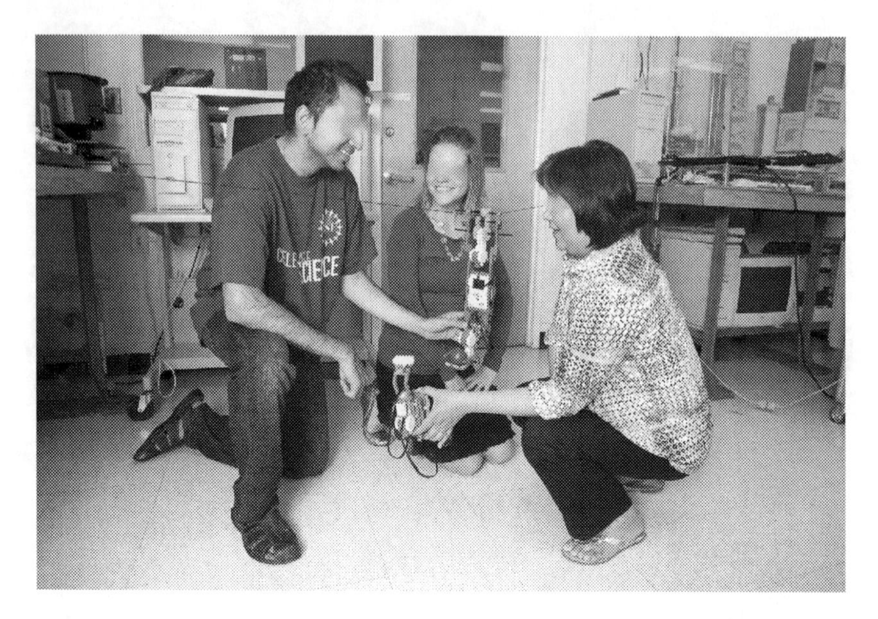

FIGURE PII.1 Creating learning experiences through educational robotics.

4

DESIGN-BASED RESEARCH FOR ROBOTICS-ENHANCED LEARNING ENVIRONMENTS

1. Introduction

Educational research is fundamental to the advancement of theories and practices that contribute to building successful learning environments. However, since it has traditionally been conducted in sterilized settings with selected groups of participants, the learning gained and knowledge produced from education research do not translate into implementable practice and thus need to be revisited for making them more trustworthy. Given the applied nature of the field, it is pivotal that education research is conducted in real-world contexts along with educators who are at the forefront of delivering instruction. In fact, researchers such as Collins [1] and Brown [2] recommend framing and conducting design experiments wherein education researchers and practitioners collaborate to generate credible results that are produced and evaluated in authentic settings. This has led to the development of design-based research (DBR), a paradigm that advocates formulating educational investigations in local contexts by partnering with practicing peers over iterative cycles of development and refinement to produce the most effective outcomes.

DBR consists of recurring steps of analysis, design, enactment, and redesign [3–5] and it is applicable in the fields of curriculum development, media and technology in education, instructional design, and teacher professional development [6, 7]. Its flexible nature allows researchers and practitioners to modify the design of DBR phases to meet the needs of the project. Moreover, its application in real contexts with input from practitioners aids in generating evolving theories that can contribute to altering educational practices [8]. In view of its contextual, collaborative, and iterative attributes, among others, the use of DBR approach in developing robotics-based lessons can help in identifying strategies

DOI: 10.4324/b23177-6

that are suitable for specific learning environments and responsive to students' needs. Moreover, this approach can catalyze student motivation and help ascertain the role of existing theories such as cognitive apprenticeship and problem-based learning in enhancing student learning outcomes [9].

To elucidate the applicability of DBR in developing robotics-enhanced learning environments, this chapter describes DBR, its characteristics, implementation processes, and examples from prior literature. Two detailed cases of using DBR to enhance the quality of learning units and a professional development program are also discussed. The chapter additionally gives an example of using systems engineering tools to improve the traditional DBR process. The outcomes of the studies and their evolving principles are discussed for each case to guide educators in designing robotics-enhanced programs for their own contexts.

2. Design-based research

A significant amount of educational research is performed in controlled settings [4, 10, 11] that can cause research findings to have limited applicability or be even inaccurate, thus leading to a credibility gap [4]. Researchers who seek to replicate and scale-up education research in a real-world context of larger and broader practice are stymied by the credibility gap, thereby failing to create an impact in the classroom [4, 11]. The experience of such research becomes further isolated due to the lack of transfer of knowledge from research to implementation practice [11]. That is, the education practitioners deem the research of the aforementioned variety to be impractical since it lacks relevance to their context and is too abstract to be useful [4]. Employing unscientific approaches to education research [4], with findings that are questionable or have limited relevance to practice [3], often yields ineffective methods to teaching and learning [2].

Having recognized the need to bridge the theory-practice gap, researchers such as Collins [1] and Brown [2] recommend that education researchers and practitioners pursue collaborative explorations. They envision that such a collaborative enterprise can help reveal how disparate learning-environment variables impact education theories and can guide and improve research implementation [1, 2]. In addition to narrowing the credibility gap, such an approach can contribute to advancing theories of learning and teaching by producing more "usable knowledge" [12] and being more "socially responsible" [13].

In this spirit, DBR represents a systematic approach that meets the aforementioned requirements and aligns with the need to enhance educational research. According to Wang and Hannafin [5], DBR is: "a systematic but flexible methodology aimed to improve educational practices through iterative analysis, design, development, and implementation, based on collaboration among researchers and practitioners in real-world settings, and leading to contextually sensitive design principles and theories."

The recursive nature of DBR aids in enhancing learning theories by using empirical evidence as a rationale for the success of a specific educational innovation [14]. Using an array of learning theories, a variety of solutions to educational challenges can be examined including educational products, processes, programs, and policies [7, 15] in such varied domains as curriculum development, media and technology in education, instructional design, and teacher professional development [6, 7]. Insights gained from the DBR methods can foster long-term partnerships, produce contextualized learning theories, and improve the capacity for education innovation [4, 5, 15, 16].

2.a. Characteristics of design-based research

DBR can be employed to conceptualize, analyze, evaluate, and improve educational interventions or artifacts for their impact on teaching and learning [4]. Unlike traditional educational research methodologies that yield siloed interpretations from educational innovations, DBR is foregrounded in realistic settings through a collaboration with practitioners and by operating in real-world settings [3–5, 17]. The implications drawn from this methodology seek to concurrently refine both research and practice [4, 17] through its iterative and flexible design methodology [3–5]. The prominent characteristics of DBR are delineated below.

DBR is conducted in authentic settings and is collaborative: Research conducted in a controlled setting, for example, a teaching or learning laboratory, generates a credibility gap by giving rise to theories that are often difficult to implement in a pragmatic setting. Alternatively, DBR is embedded in authentic educational contexts, and it is conducted under a collaborative partnership between education researchers and practitioners, for example, curriculum developers and teachers [4, 5, 17]. DBR is often performed for an extended period in the same setting [4, 5] to facilitate recursive refinements to the designed innovation. Such a methodology provides a framework for examining the validity of research results, which in turn can effectively improve educational practice in comparable contexts [17]. It is critical to include practitioners as participants in the DBR methodology. The practitioners bring varied levels of expertise to the research endeavor and can contribute to identifying the problem and creating design principles [17]. Moreover, as collaborators, they can help in executing the research plan and analyzing its outcomes [8]. In such a partnership, both researchers and practitioners learn from each other [15] and co-create authentic design principles that can be put to broader use.

DBR is iterative in nature and refines both theory and practice: DBR seeks to uncover the complex mechanisms that affect teaching and learning in realistic situations [15]. In support of this aim, the research is developed and performed through recurring steps of analysis, design, enactment, and redesign [3–5]. The flexible nature of the plan allows the researchers to make thoughtful changes in response to the obstacles encountered [5, 15]. Iterative refinements to the enacted plan and the resulting findings are examined and documented within each implementation cycle [4, 15]. Additionally, DBR is based on preliminary ("proto") theories and propositions that are refined or generate other discernible patterns, which can enhance the proposed design [4, 15, 17]. Contextual application, input from the practitioners, and detailed analysis all contribute to advancing theories [4, 17]. It is imperative that emerging theories impact similar contextual settings, communicate relevant implications to other educators, and lead to a significant shift in educational practice, eventually [8].

DBR uses mixed research methods: Since DBR focuses on improving an educational intervention, its research methods can vary as emerging iterations necessitate different focus. Thus, mixed research methods are adopted to establish the reliability of ongoing research [5]. Each research phase might have a differing requirement and, hence, the designer can select the method deemed to be the most suitable and rigorous in yielding credible results [17]. While the procedures are flexible, they are not truly undefined [8]. An initial tentative research design is revised per the research findings and, thus, a "black-box model" of measurement is discouraged [3].

At a first glance, DBR may seem to have close resemblance to other education research methods such as formative evaluation, due to commonalities in their iterative nature, or action research, because of similarities in being placed in authentic contexts and collaboration with practitioners [5]. Despite such surface resemblances, DBR is evidently distinct from these other approaches due to its theory-driven nature, wherein the research is grounded in existing theories that are tested and evolved to generate new theories [5]. DBR does not replace the other approaches but rather incorporates them to provide improvements and reliability in research [5]. This means that other research methods used in tandem with DBR on a case-by-case basis can lead to more robust results. Table A.2 in Appendix A summarizes the differences between DBR and several other methods, namely, laboratory research, randomized control studies, formative evaluation, and action research.

2.b. Design-based research process

The traditional approach of DBR encapsulates the core processes involved in most design-based interventions and follows continuous cycles of iterative development. Specifically, as summarized in [18], the process of DBR begins with identifying a relevant problem and creating a prospective solution that could be in the form of a curriculum, an educational product, a teacher professional development plan, an instructional methodology, etc. Next, the prospective solution is field-tested in partnership with practitioners and other participants to observe its impact. Then, the intervention is evaluated for its effectiveness through comprehensive data collection and analysis, and the design of the intervention is enhanced through iterative refinements. Finally, through reflection, various aspects of the intervention are examined to ascertain their impact, and appropriate amendments are made. Having implemented these stages and depending upon the project needs, additional cycles of iterations may be pursued to further refine and improve the overall outcome of the intended solution. While this is an approximate model, the sequence of events may change and not all activities may be performed, especially in later iterations. McKenney and Reeves [19] have delineated in detail the following core processes and their encompassing activities that represent the traditional approach of DBR (see Figure 4.1).

Analysis and exploration: The primary aim of this phase is to improve the definition of the problem and to generate initial design requirements that can aid in addressing the problem. To do so, this phase encompasses collaboration with practitioners to seek, gain, and refine the comprehension of the problem. Furthermore, literature review serves as a critical component in developing an understanding of the feasibility of attaining the solution and in guiding data collection and analysis efforts to derive meaning from the outcomes. Alongside this initial analysis phase, the exploration phase seeks a deeper understanding of the problem and collects insights from experts and practitioners who may even contribute to collaborative brainstorming.

Design and construction: The goal of this phase is to refine the potential solution, goals, and design requirements using existing and robust theories, informed and guided by local expertise. It also focuses on engendering novel solutions and assessing their viability. The most promising solutions are enumerated and mapped to the specified design requirements. Next, the mapped solution list undergoes prototyping where partial but functional designs are created and sometimes tested. This process may lead to the actual representation of the final design intervention itself.

Evaluation and reflection: Once the artifacts for intervention have been constructed, they are deployed in authentic settings to check for their appropriateness and viability. Implementation is often accompanied by or followed by the *evaluation phase*. Evaluation refers to an empirical analysis of the mapped solution or prototype using both qualitative and quantitative methods. In the *reflection phase*, which occurs following evaluation, researchers and practitioners thoughtfully study the outcomes with the aim of revising current methods to ultimately produce a refined version of the intervention or evolve design principles. Evaluation and reflection are paramount for all design interventions and are encouraged informally throughout all phases.

Note: Although the core processes of DBR are listed here sequentially, they may occur simultaneously or in a different order, as required in a study.

2.c. Contributions of design-based research

The DBR methods can contribute to advancing both educational theory and educational practice since they are situated in authentic settings, grounded in

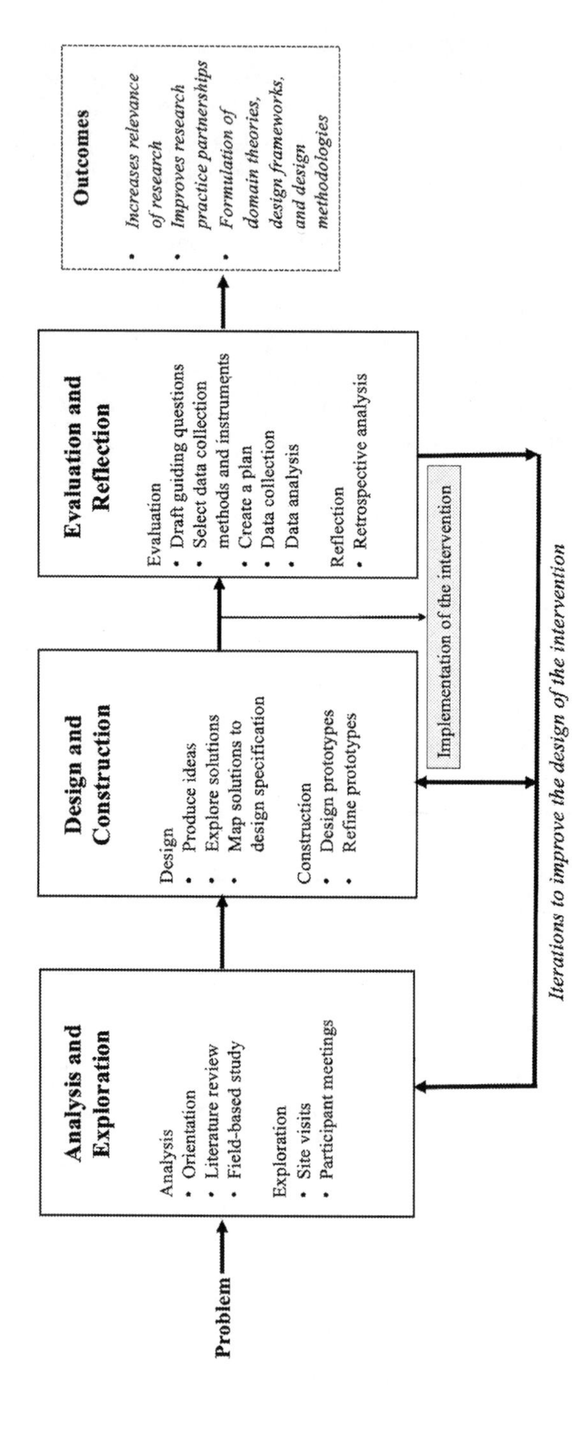

FIGURE 4.1 Traditional approach for implementing design-based research (as compiled by McKenney and Reeves [19]).

theories, and implemented in collaboration with practitioners. Practical contributions of DBR have primarily been in the fields of curriculum development, media and technology in education, instructional design, and teacher professional development [6, 7] as well as in validating educational products, processes, programs, and policies [7].

Since DBR is conducted in real-world settings, researchers are afforded the opportunity to understand and test the educational intervention as well as amend and enhance their designs based on the contextual challenges that they uncover. Applying research principles in authentic scenarios helps in filling the "credibility gap" and in generating knowledge that may be germane in similar contextual settings. Results of such research have greater potential to influence the design of other educational interventions and can be reliably adopted in educational practice [4, 5, 15]. Furthermore, research that is conducted in partnership with practitioners and that addresses their genuine interests, needs, and curiosities, generates findings that have greater utility to them and are readily implementable in their classrooms. Thus, DBR helps in contributing to research that is useful beyond research settings [4, 15]. The collaborative interactions with practitioners foster perspective-taking among researchers to better understand variables in naturalistic contexts that lead to changes and iterative improvements to designs, helping transfer knowledge from practice to theoretical research [4, 5, 16].

DBR can contribute to the development of contextualized theories of learning and teaching that help in explaining phenomena, predicting causal relationships between selected variables, and recommending strategies that could lead to improved practices [7]. Edelson [16] suggests that, unlike a linear research approach that is oriented toward evaluation, the DBR paradigm provides avenues to researchers to understand the teaching and learning systems and helps in formulating three different types of theories. First, *domain theories* suggest opportunities and challenges of learning environments and their contribution to learning as well as findings pertaining to teacher and learner requirements. Domain theories are descriptive in nature, indifferent to research design, and contribute to developing ideas that may impact other educational interventions. For example, an instructional designer may uncover specific challenges related to utilizing a software-based environment with learners [16]. Second, *design frameworks* are prescriptive in nature and provide a generalized and coherent set of guidelines that the designed artifact must possess to be successful. For instance, anchored instruction is a form of a design framework that emphasizes meaningful problem-solving [16]. Third, *design methodologies* are general procedures that explain the process to achieve the desired design results and the role that various experts play in attaining them. Design methodologies are utilized to measure the progress and address the challenges encountered during design implementation. For example, a design intervention incorporates methodologies to collect user feedback and incorporate it in future design iterations [16].

3. Literature review exemplifying the use of design-based research in robotics-enabled learning

The principal attribute of DBR is its recursive nature that contributes to improvement in educational interventions and supports theory-building for wider use by the education community. While DBR is initiated with certain planned processes, its flexible nature allows the designers to alter the design in response to contextual needs. Moreover, the DBR process undergoes cyclic steps of analysis, design, enactment, and redesign to best meet real-world requirements. Thus, different designers may select varied processes to obtain the desired outcomes. This section presents two research studies that relied on the use of DBR processes (as summarized in Subsection 2.b) and that serve as examples of collaborative research in authentic settings for enhancing the learning theories and artifacts.

A design-based research study [20], situated in Norway, intended to refine a constructionism-based coding workshop for participants aged 8–17 years. The workshop sought to investigate students' learning experience in coding activities, which used a blocks-based programming tool and digital robots to create meaningful artifacts in the form of games. Among the other goals of the study, DBR was used to improve the engagement, scalability, and sustainability of the workshop and to determine the emerging principles that can add value to similar constructionist activities. The workshop was improved iteratively over three DBR cycles with a total of 157 participants (101 males and 56 females). Each iterative cycle had a different set of participants and employed mixed methods of data collection and analysis.

For the first two cycles, a four-hour workshop guided students to: interact with digital robots using simple coding concepts such as loops, outline storyboards and visualize their own games, code their own games using the blocks-based programming tool collaboratively, and reflect on games developed by other teams. As per DBR, key findings from one cycle of the workshop informed the design and execution for the next cycle. For example, some of the key findings of cycle two were to: allocate sufficient duration of time for students to collaborate with others on their team, get acquainted with the programming tool, and express their creativity. One of these findings led to an increase in the workshop duration by making it a two-day endeavor spanning for five-hours each day. Other findings of various cycles suggested several refinements to the workshop, for example, the first implementation cycle suggested utilization of a scaffolding approach to reduce cognitive load, and the third cycle recommended demonstration of illustrative examples of female game heroes to reduce the stereotype that only boys like video games. Multiple recommendations for each cycle also informed the formulation of several design principles that can facilitate student learning for constructionism-based coding activities including engendering social interactions in learning, designing meaningful learning activities, fostering student motivation, etc.

Yet another study [21], situated in the United States (US), employed DBR to improve an *integrative* science, technology, engineering, and mathematics (STEM) curriculum that sought to develop students' computational thinking abilities with the use of robots. The integrative curriculum aimed to draw on content from multiple STEM disciplines to engender student interest and teacher acceptance. Applying the conceptual framework of the DBR process, during the first phase, the researchers analyzed and explored literature and design principles that were fundamental to the design of an effective integrative STEM curriculum. They identified three factors critical to the design of their curriculum, namely, the creation of a functional environment that allows authentic problem-solving, inclusion of opportunities for hands-on embodied learning, and integration of STEM standards in the curriculum. During the second phase of DBR, the researchers designed a curriculum based on the design principles from the first phase. The robotics-based integrative STEM curriculum catered to a fifth-grade classroom and consisted of six lessons that would take ten 50-min periods to complete. Each lesson included an authentic and open-ended science problem and required students to be guided through constructing and programming robots to discover prospective solutions.

During the final phase, eight fifth-grade teachers participated in a professional development program and were provided with a teacher's guide for ease of implementation. Then, they implemented the curriculum in their classrooms that served a total of 263 students. For both teachers and students, data was collected and analyzed on the basis of the three design principles from the first phase. The results were used to formulate recommendations that would allow improvements to the curriculum design for the next cycle of implementation. For example, two of the key recommendations included the need for additional support in robot building and clearer guidelines in the teacher's guide. Thus, for the next cycle of implementation, it was decided to provide additional in-class support for implementation and provision of easy-to-use instructional preparation material such as PowerPoint presentations and videos in addition to a simpler teacher's guide.

The implications of outcomes from both the examples are representative of the effective contributions that the DBR approach can make in improving learning artifacts and generating principles that can be used in other contexts.

4. Design-based research implementation examples from robotics-enhanced learning environments

Educational technology interventions are often founded upon incompatible and confounding theories that lead to inconsistencies between their implementation plans and practical realities. DBR helps in addressing this gap between theoretical underpinnings and practical usage of such technology-based educational innovations. Although a typical instructional design cycle evaluates the effectiveness of the intervention, it does so upon its completion rather than during execution. Thus, DBR is evidently more suitable for designing and enacting technology-enhanced learning environments [5].

Using robotics to inculcate technology-enabled learning experiences can contribute to student achievement by promoting active participation, imparting improved comprehension of abstract topics, catalyzing greater motivation in learning, and creating a better learning environment [22]. Implementing DBR to design robotics-enhanced lessons can help in identifying the most reliable and feasible approaches in promoting student motivation given the flexible nature of robotics tools [9]. Additionally, using DBR to design STEM-integrated robotics lessons can yield useful information on the suitability of various learning theories such as cognitive apprenticeship, situated cognition, problem-based learning, and inquiry-based learning, among others [9].

This section comprises three DBR scenarios that were implemented in actual classrooms to enhance the quality of robotics-based STEM lessons or professional development programs and to suggest evolving principles that can be adopted by the wider education community. These studies were conducted by a collaborative team of engineering and education researchers and middle school teachers from urban schools. The first and the second examples employ traditional DBR processes to iteratively develop a lesson and a professional development program, respectively. The third example showcases the use of systems engineering tools and practices along with the traditional DBR approach to refine a lesson design, leading to improved student learning outcomes.

4.a. Example 1—Iterative improvement of robotics-based learning sequence in middle school [14]

This DBR activity was implemented in three seventh-grade classes with 77 students (46 males and 31 females) and it was conducted by the engineering and education researchers in collaboration with one teacher. The researchers aimed at developing theories that could be applied in practice to increase student motivation toward STEM subjects. This particular study aimed at maximizing student learning by evolving constructionism theories [23] through robotics-enabled learning environments along with several other theoretical constructs such as problem-based learning [24], cognitive apprenticeship [25, 26], and anchored instruction [27]. It was conducted over several months to give adequate time for making each iterative refinement. The same teacher and students participated in all the iterations.

(i) Theories and framework

First, constructionism is a popular learning theory that encourages learners to build their own knowledge by creating tangible artifacts [23]. This theory

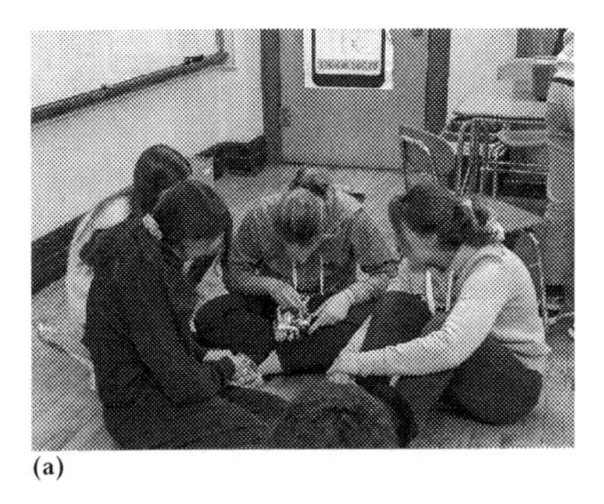

(a)

(b)

FIGURES 4.2 (a)–(b) Students engaged in constructing artifacts for robotics explorations.

naturally aligns with robotics-based learning wherein students build, program, and operate robots to derive meaning and construct understanding for the material being taught and learned. Thus, constructionism was selected consciously as the foundational theory for this study [14]. Second, problem-based learning allows students to grasp concepts and build thinking skills by solving a relevant and authentic problem [28]. Third, cognitive apprenticeship supports learners to integrate the learned content into real-world contexts [25]. Fourth, anchored instruction places student learning within a relevant, problem-solving context [24–27]. The robotics learning sequence, as first developed by the researchers, consists of the following five phases.

- Robot chassis—understanding the design and construction of a robot as well as learning about frames, symmetry, load, and center of mass.
- Drive mechanism—understanding about gear ratios and motors through lessons.
- Transducers—learning about the different sensors and actuators.
- Robot motion—understanding about translation and rotation in a robot.
- Programming—programming different robots as per different tasks.

Throughout the aforementioned learning sequence, students were to be engaged in myriad tasks to design, construct, integrate, program, test, and iteratively enhance their robotics creations. See Figures 4.2 (a)–(b) illustrating students engaged in constructing robotics artifacts.

(ii) Iterative design interventions

DBR was employed to improve the design of the first three phases of robotics learning sequence (i.e., chassis, drive mechanism, and transducers) and to develop usable theoretical knowledge that could be adopted by other educators. The traditional DBR process of Subsection 2.b, with adjustments according to the needs of the research, was used in five iterative cycles focused on the three phases of the robotics learning sequence. The DBR process is launched with the following considerations that align with the analysis and exploration step of the process. To judge the effectiveness of any technology intervention, Ferdig [29] suggests the following assessment criteria—appropriate use of technologies, content learning outcomes, and deep analysis of affective (social and emotional) and cognitive gains (see also Chapter 6). In [14], the first and third criteria of Ferdig were adopted to revise the learning design using the data collected primarily from classroom observations, pilot pre- and post-tests, and student work artifacts. To inform the design of robotics learning sequence, prior to the introduction of robotics-related instruction, a researcher visited the three classrooms of the collaborating teacher to observe the learning environment, student interactions, and instructional processes. While it was observed that the students were disengaged and disruptive during the traditional didactic instruction, the introduction of computer-aided instruction reduced student interruptions and allowed them to work at their own pace. Based on this initial understanding, it was determined that organizing students in small teams to stimulate a social learning environment and involving students in their own learning process through activities can engender their motivation to learn. Table 4.1 delineates iterations for the robotics learning sequence.

(iii) Outcomes of the study [14]

This study led the research team to reflect on aspects of the design that helped in achieving the goals and ones that did not. Classroom observations showed that the students responded to learning the most in the presence of a relevant and engaging problem, for example, creating a ball sorting mechanism or a zombie escape machine. The pre- and post-test data for the gear ratio lesson further reveals the positive impact of relevant problem-based learning as showcased by an improvement from 35% average pretest score to 72% average post-test score.

Embedded with the constructionism learning theory and problem-based learning approach, robotics-enabled learning was quite successful in engaging and retaining student attention. The presence of constructionism in the form of building their own robot configurations led the learners to readily surface and address misconceptions, learn through their peers, and ultimately, create their own understanding. The use of problem-based learning gave opportunities to students to conceptualize solutions and the integration of cognitive apprenticeship guided them toward the solution. The application of anchored instruction was deemed unfit within the context of DBR as it requires a high level of details that are challenging to revise with each upcoming iteration within the DBR cycle. Lastly, the learning sequence also catered to a variety of learning needs, for instance, the PowerPoint presentations helped in facilitating visual learning, classroom lectures and discussions reached auditory learners, and robotics-based activities supported kinesthetic learners.

TABLE 4.1 Iterative improvement of a robotics-based learning sequence in middle school setting [14]

	First iteration—Robot chassis design
Learning sequence (Design and construction phases)	The teacher launched the instructional sequence with a whole-class instruction, introducing the uses of robots in daily lives and the relevance of STEM disciplines to creating robotics solutions. Next, the students were shown a video of a robot solving a Rubik's Cube, which was followed with an introduction to and discussion on various frame structures for robots. Following this initial introduction to robots, under the framework of problem-based learning, the students were presented with a LEGO robotics problem: *How do we create a robotics apparatus that sorts blue and red balls contained in a tube into two containers?* Using the insights gained from initial classroom observations and suggested recommendations, the students were organized in groups where they identified problem constraints and solution criteria, such as materials, time, sensors, etc. Finally, the students were encouraged to use LEGO robotics kits to build chassis prototypes on their own.
Class observations (Evaluation and reflection phases)	While the students were engaged with the initial video, they were inattentive to the introductory lecture, which precipitated in disruptive behavior. Many students struggled to build practical design of chassis, especially for two reasons—some were perplexed with the large number of LEGO pieces and some did not have prior building experience with LEGO, which prevented them from effectively connecting the components together.
Recommendations for next iteration	Based on these observations, four recommendations were devised to address student disengagement and to increase their involvement in the design of a robot chassis. First, the teacher suggested that the use of worksheets can reduce interruptions and, thus, several worksheets were introduced in the next iteration. Second, the lecture on robot frames was divided into several small portions, each interspersed with worksheet activities to keep the learners more engaged. Third, the teamwork activity for creating chassis designs was altered from open-ended construction tasks to the students sketching and clarifying their designs first. Fourth, to cater to the range of prior exposure with LEGO, additional instruction on designing and building using the LEGO material was offered to the students.

(*Continued*)

TABLE 4.1 (Continued)

	Second iteration—Robot chassis design
Learning sequence	Recommendations from the first iteration of the instructional sequence were employed to revise the lesson for the robot chassis design phase. Worksheets were provided to engage the students and help them in visualizing the concepts. As opposed to a small-group discussion in the first iteration, the teacher led a whole class interaction to identify constraints and criteria for designing the robot. After building an initial understanding of robot design, the students worked in groups to brainstorm and sketch possible robot designs and engaged with other groups to compare design ideas. Under cognitive apprenticeship, the researcher, who is a mechanical engineer, modeled a design to help the students in comparing their designs with a working example. Finally, using step-by-step instructions and through teacher assistance, student groups built their own LEGO robot designs.
Class observations	Greater engagement and reduced disruptive behavior by the students were observed during this revised learning sequence. The revised sequence met Ferdig's first criteria concerning appropriate technology integration, as the lesson included opportunities to showcase student learning and to apply it in building the chassis. Moreover, the sequence met Ferdig's third criteria concerning affective gains, by incorporating social aspects. Finally, cognitive apprenticeship, in the form of comparison with an expert's design, promoted student reflection for their own problem-solving for the ball sorting activity.
Recommendations for next iteration	The research team decided to introduce anchored instruction in the next iteration to have the students explore the problem more closely. The students were prepared to move from the design of chassis to drive mechanism in the robotics learning sequence.
	Third iteration—Drive mechanism
Learning sequence	The learning sequence advanced to robot drive mechanism and anchored instruction was utilized for the lesson. A PowerPoint presentation was used by the instructor to introduce gear ratios, inputs and outputs of gears, and their relationship to robot speed. A scenario was introduced: *Create a robot to escape from a zombie horde using a gear train that closes the door behind which the students can hide during a zombie apocalypse. The robot can run for only 3 seconds at a time and you have a limited collection of gears. As a team, you will determine which gear ratio will allow the robot to move the farthest in 3 seconds and record observations using a worksheet.*

Class observations	While the students stayed engaged in the activity, the zombie theme appeared distracting for them as they got involved in alternative discussions. It was hard to establish whether the students were engaged because of anchored instruction or the need for a particular solution to the problem. It was also found that most students were able to identify different workable gear ratios and only some of the groups struggled. Finally, some students were unable to collect data systematically despite guidance from the worksheet.
Recommendations for next iteration	The observations revealed that anchored instruction was ill-suited for the DBR framework as anchored instruction requires a greater level of detail and planning for which multiple changes within the DBR iterations are challenging to execute. Moreover, reduction of zombie imagery was suggested in the upcoming iteration due to the distractions caused by it. Finally, it was recommended to explain the gear testing section in more detail by breaking it down into specific steps to provide greater support for student learning.
	Fourth iteration—Drive mechanism
Learning sequence	In this iteration, the students were provided with scaffolded descriptions of gear ratios, their effect on the robot speeds, and the test process. They completed a worksheet that included problems based on more complex gear ratios and provided their reflections to a questionnaire at the end of the worksheet.
Class observations	The pace of student comprehension of gear ratios and their ability to test gear trains had notably increased, which met Ferdig's first criteria concerning appropriate technology integration. The act of scaffolding for the gear ratios and testing sections allowed the previously struggling students to have an increased engagement as well. In discussions and from responses to the questionnaire, an expansion in the student use of engineering vocabulary words was noticed. An analysis of the reflection questionnaire revealed that the students were able to relate their learning to a math lesson on ratios, which indicates the accomplishment of Ferdig's third criteria concerning cognitive gains.
Recommendations for next iteration	Having observed an increased use of engineering vocabulary among the students, the research team suggested including a vocabulary section on the worksheets to help them internalize the meaning of new words and be able to verbalize explanations. The reduction of zombie imagery did not lower student motivation in learning and, thus, it was eliminated from future lessons. It was further decided to focus on the central problem of ball sorting for the rest of the robotics learning sequence.

(*Continued*)

TABLE 4.1 (Continued)

	Fifth iteration—Transducers
Learning sequence	In this phase, the teacher used a video and a worksheet to explain color sorting through real-life applications and asked the students to identify various sensors in the video and their uses. Next, the teacher discussed different types of sensors in the LEGO robotics kit, how light sensors work, and the principle of reflectivity. The students were divided in groups to brainstorm ideas about how balls can be sorted by their colors using measurements from a LEGO robot's LCD screen. The students were provided with blue and red balls to record the amount of light reflected. Finally, based on their learning from the previous lessons and experiments, the students ordered different colors based on their reflectivity.
Class observations	It was observed that the students used the vocabulary terms effectively during this phase and were able to make connections of what they were learning in this lesson to previous lessons, meeting Ferdig's third criteria concerning cognitive gains.

(iv) Evolving principles [14]

Ongoing cycles of design and development for this robotics-based learning sequence led to the evolution of several principles that can be employed in other settings that use robotics as a learning tool. Some of these principles are delineated below.

1. In designing a new robotics-based learning sequence, teachers must be well-acquainted with students' prior knowledge and skills, especially pertaining to the robotics components, sensors, programming, etc. In the design of [14], the assumption that students had prior exposure to LEGO components or construction was incorrect as they encountered challenges in recognizing components and building prototypes for the robot chassis activity in the first iteration.
2. Splitting lessons into manageable, bite-sized, focused portions can be more effective. Moreover, when feasible, interspersing lectures with worksheet activities or hands-on experiments with robots can engage learners and help them keep pace with the teacher. It was seen that splitting the lesson into smaller parts in the robot chassis construction activity led to better engagement and providing worksheets increased learning effectiveness.
3. Beginning a lesson with whole-class instruction and discussion and subsequently proceeding to small group activities can be more beneficial than solely relying on small group discussions. For example, in the iteration

launched with the small group setup, it was observed that inter-group disruptions limited the participation of even the enthusiastic students. To remedy this, in the next iteration, a whole-class discussion was instigated prior to the introduction of small group activities, and this reorganization led to more productive outcomes.

4. Learning sequences with high-excitement storylines can potentially be distracting to students. Thus, it is better to forego such stories, especially if they fail to add value to student learning and are instead used primarily to draw student attention. For instance, the zombie imagery led students to discuss unnecessary details that distracted them from solving the problem.
5. Students' prior knowledge must be activated to give them opportunities to organize knowledge, surface misconceptions, strategize solutions, and be more receptive to new learning experiences. If feasible, opportunities for cognitive apprenticeship such as modeling for real-world challenges can help learners relate with the problem and become more motivated to pursue STEM fields. For instance, in problem-based learning experiences, such as the one described in this design, students can learn a lot from mistakes and the directions they take to reach a solution, as opposed to merely obtaining a result.

4.b. Example 2—Iterative refinement of a teacher professional development program [30]

There is an increased demand for professional development offerings that can equip teachers with capacity to incorporate new technologies in their practice and prepare students for success in a world of increasing technology. This subsection describes a robotics-focused pilot teacher professional development program that was refined using the methods of DBR. Four teacher participants, two science and two math teachers, collaborated with the engineering and education researchers over a duration of three weeks. The primary goal of this research was to assess the role of situated learning [31] in professional development. The study also sought to gauge the impact of the professional development program on teachers' science and math content knowledge and their skills to design robotics-based lessons. Finally, the study examined how DBR can help refine the learning plans and teacher professional development program created by the researchers.

(i) Theories and framework

Most educational technology-focused teacher professional development programs are successful when they: utilize hands-on activities to engage teachers as learners, employ constructivist approach of instruction, and offer continuing support to teachers for a year [32]. Along with the incorporation of these features, the professional development sessions in the study of [30] were designed in the context of situated learning [31], which refers to collaboratively creating meaning by participating in real and authentic activities. In this case, the collective group of educators focused on building standards-aligned robotics-based lessons and embedding robotics in their pedagogy as opposed to using it simply as a novel playful tool.

To fully engage the teachers and to introduce them to new knowledge and mechanics of robotics-enabled teaching and learning experiences, the professional development workshop of [30] was consciously designed and consisted of three modules. Module 1 (*Introduction to LEGO robotics kits*) focused on familiarizing the teachers with different components in the kit including sensors and the process of assembling a base robot by following the given directions. Module 2 (*Collaborative group work in refining standards-aligned lessons*) introduced the teachers to sample lessons created by the researchers, allowing them to work through the lessons as learners to examine the relevance, opportunities, and challenges of robotics activities for curriculum, classroom, and students, and encouraging them to enhance lesson and activity designs for tighter alignment with standards. Module 3 (*Review and discussion of education research literature*) introduced the teachers to topics such as technological, pedagogical, and content knowledge (TPACK) (see Chapter 6); Design-based research; problem and project-based learning; cognitive apprenticeship; and anchored instruction.

(ii) Iterative design interventions [30]

DBR was employed to make two levels of iterations in this program: refinement of robotics-enhanced science and math lessons by the teachers and researchers, collectively, and improvement in the design of professional development for the teachers who are expected to deploy the resulting lessons in their classrooms. The traditional DBR process (see Subsection 2.b), with contextual adjustments, was employed to inform these iterations. Table 4.2 describes the evolution of the program over three weeks of the professional development program. Figure 4.3 illustrates teachers testing a robotics-based lesson.

(iii) Outcomes of the study [30]

After the completion of professional development, the teachers were visited and observed in their science and math classrooms. They were found to have gained proficiency in loading various LEGO programs for operating robots and identifying programming difficulties. This shows, in the context of situated learning, that the teachers had moved from the "periphery of the robotics community to the local novice community of practice centered on utilizing robotics in their science and math lessons." Moreover, in pre- and post-tests for robotics-related science and math content knowledge, the teachers demonstrated averages of 46.4% mastery prior to professional development and 61.6% mastery after the professional

TABLE 4.2 Iterative improvement of a teacher professional development program through situated learning techniques [30]

	Week 1
Learning sequence	The first week of the program began with all three learning modules. Module 1 included introductory lectures and hands-on activities to familiarize teachers with the components and programming environment of the LEGO robotics kit. Module 2 introduced the teachers to three robotics-based lessons on modeling, energy, and expressions and equations through PowerPoint presentations and related hands-on robotics activities to facilitate situated learning. For the hands-on activities, instructions were included on robot construction and assembly, allowing each two-teacher team personalized pacing. As each learning unit of Modules 1 and 2 was introduced to the teachers, they worked through its steps using corresponding worksheets from a student perspective. Finally, under Module 3, the teachers were provided with research readings on TPACK and prompted to engage in a discussion on it.
Observations	When performing the hands-on learning activities of Module 2, the teachers found it challenging to follow the robot build instructions as they did not deem them intuitive. Moreover, they found the student worksheets for lessons to be lacking a natural flow and alignment with the activities and thus they deemed them hard to follow. The teachers and researchers had to redesign the base robot to produce an artifact that can be used across all planned robotics-based lessons and activities.
Recommendations for next iteration	To improve the flow of the professional development program, a new computer-aided design tool was suggested for producing detailed and clear build instructions for the robotics platform. The student worksheets for each lesson required modifications according to feedback from the teachers. Finally, for the next iteration of the professional development program, it was recommended to increase teacher involvement in lesson development to reflect student perspective.
	Week 2
Learning sequence	During the second week, under Module 2, the teachers were introduced to four robotics-based lessons on ratios and proportions, analyzing and interpreting data, statistics, and functions. In addition to performing hands-on learning activities, now the teachers were engaged in developing relevant assessment material for each lesson. Moreover, for Module 3, the teachers participated in reading articles on the topic of design-based research and were prompted to engage in a discussion on it.

(*Continued*)

TABLE 4.2 (Continued)

Observations	The teachers found it challenging to navigate certain activities with the LEGO robot kits such as creating plots with multiple data sets. They also deemed some lessons lacking due to their failure to involve students in various aspects since the software did most of the work from acquiring to plotting the data. In this way, the teachers began to critically think of robotics as a pedagogical tool and became more proficient in using robots.
Recommendations for next iteration	It was suggested that the revised professional development program should increase teachers' interaction time with the LEGO robots to improve their ability and to provide them more opportunities to develop lessons on their own. It was recommended that robotics activities embed opportunities for students to plot data by hand as opposed to merely using the LEGO software for doing such work.
	Week 3
Learning sequence	Based on their learning and exposure from the first two weeks, under Module 2, the teachers developed new lesson plans, activities, and assessments using two different approaches. The first approach was a content-driven approach where the teachers identified a topic that students find challenging to understand and used robotics to cater to those concepts. The second approach was a technocentric approach where the teachers were prompted to use a robot sensor, which had not been introduced to them yet, for teaching a STEM topic. Alongside the development of new learning material, for Module 3, the teachers were involved in research readings on intrinsic and extrinsic motivation, project-based learning, cognitive apprenticeship, and anchored instruction. Toward the end of the program, the teachers presented their learning.
Observations	In the lesson development phase, the teachers brainstormed challenging topics collectively (such as addition and subtraction of integers) and created lesson plans and corresponding robotics activities. While the teachers desired for the robots to work in a particular manner, they lacked the abilities to meet their goals as they found it difficult to navigate the programming interface.
Recommendations for next iteration	In the revised professional development program design, it was recommended that scaffolding be incorporated during the exposure to the LEGO robotics programming interface with a step-by-step instructional guide. It was also recommended that a community of practice be formed to continue the collaborative support and discussions among the teachers.

FIGURE 4.3 Teachers engaged in testing a robotics color sensor.

development, yielding a 32.8% increase. This professional development met the objectives highlighted earlier by showcasing positive effect on teacher content knowledge, improved ability in developing and implementing lessons using LEGO robotics, and enhancing robotics-based science and math lessons developed by the researchers.

(iv) Evolving principles [30]

Integrating new technology in a classroom can be a challenging task for teachers, especially new teachers, and an effective professional development can help in alleviating this pressure. The study revealed the following evolving principles that can guide the design of an effective robotics-focused professional development program.

1. Feedback by participating teachers on various aspects of the design and implementation of the professional development program, and their ongoing recommendations for its improvements, can contribute to the growth of teachers themselves. For example, this DBR-guided teacher professional development program was effective largely due to on-going feedback from teachers. The researchers purposely created space and time for the teachers to offer feedback for improving the professional development program and suggest future iterations for creating high-quality lessons and activities.
2. Teachers must be provided with a scaffolded approach and sufficient time to explore various construction blocks, sensors, motors, and programming interface related to the robotics kit. For example, in the PD program when

the teachers were afforded hands-on opportunities to explore various components in the robotics kit and step-by-step instructions for programming, they developed proficiency in using robotics as a pedagogical tool and created relevant lessons.

3. After being exposed to robotics fundamentals (e.g., components, construction, programming, and relevant literature), based on their classroom needs, teachers can be offered opportunities to develop their own lessons to help them gain confidence in practically applying their knowledge. For example, in the case of this PD program, teachers gradually gained fluency in developing their TPACK (see Chapter 6) and, thus, in designing pedagogically-sound robotics-based learning units.
4. Creating a community of practice that can provide on-going feedback and learning support beyond the duration of professional development can be beneficial for the participants. For example, by the end of the aforementioned PD program, a community of practice had developed that fostered sustained opportunities for learning, discussions, and implementation support.

4.c. Example 3—Iterative improvements using systems engineering tools and practices [9]

Situating DBR in local settings is the principal reason that leads to reliability in its research results. While the outcomes achieved through a DBR implementation are useful and replicable for that context, they may not be generalizable to other settings and thus may be incapable of supporting large-scale implementation [8, 9]. Moreover, the flexible nature of DBR can pose difficulty in identifying causality, that is, credibly recognizing, isolating, and analyzing the factors that significantly impact the emerging outcomes in the form of theories and principles [4, 9]. This may produce gaps in suggesting appropriate revisions to the next iteration, thus limiting the overall effectiveness of DBR. Recent research [9] on developing a robotics-based STEM lesson through DBR has curated a set of systems engineering tools and practices that can be applied in tandem with the traditional DBR processes. Such an approach can help overcome the challenge of identifying causality and thus strengthen the quality of iterative refinements.

(i) Theories and framework

In the approach of [9], DBR is treated analogously to the continual improvement process (CIP) [33]. As the name suggests, CIP is an on-going enhancement in the performance of products through gradual or radical improvements. Plan–do–check–act (PDCA) [33, 34] is an iterative four-step method that supports CIP (see Figure 4.4).

After the identification of a problem in need of CIP, the *plan* phase specifies desired goals that establish the success of a proposed intervention and lists the tasks

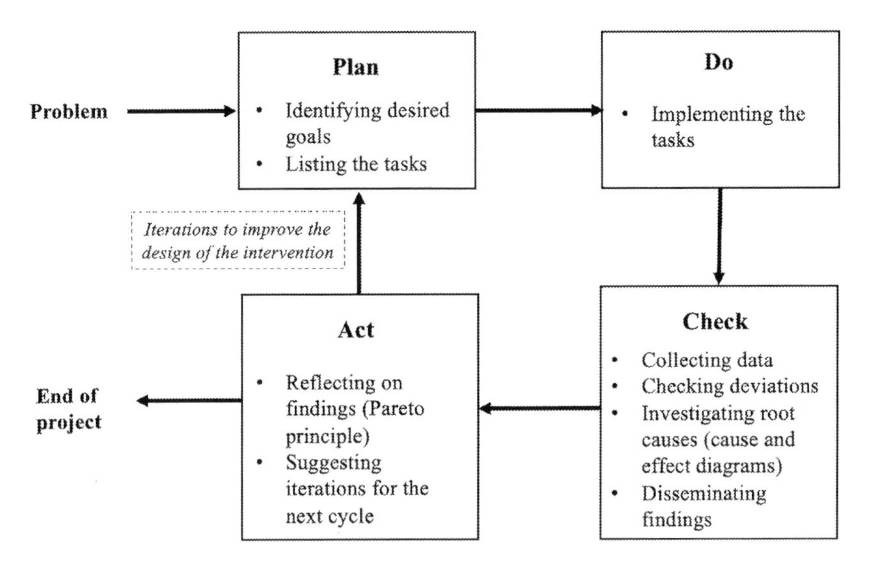

FIGURE 4.4 PDCA cycle to enhance the effectiveness of DBR.

that will aid in accomplishing the goals. The *do* phase entails implementing the tasks aligned with the goals of the *plan* phase. The *check* phase calls for acquiring the performance data and examining it against the specified performance goals from the *plan* phase to determine any deficiencies. Furthermore, this phase also involves examining for deviations in planned tasks *versus* actual implementation, investigating the root causes, and using graphing techniques to disseminate findings. *Cause-and-effect diagrams* [33], also called fishbone diagrams, can be useful to ascertain probable reasons behind the unwanted outcomes of the intervention. These diagrams are created by identifying categories that have led to the unwanted result and, then, establishing causes behind each of those categories. Finally, the *act* phase includes reflection on the root causes and findings and recommended iterations for the next PDCA cycle. Once the causes are identified, they are analyzed using the *Pareto principle* [35]. The Pareto principle, popularly called the 80:20 rule, suggests that a large number of effects are caused by a small number of factors, or, 80% of the outcomes are caused by 20% of reasons. This principle can help in identifying the few crucial factors that impact the majority of the unwanted outcomes and, hence, in suggesting changes for future iterations. These 20% of factors are improvement inputs determined in the *act* phase for the next PDCA cycle.

Through the DBR-focused research in a robotics-enhanced STEM classroom, the study of [9] showcased that embedding a systems engineering approach in the DBR framework can produce more effective iterations. A robotics-based lesson was implemented in a seventh-grade classroom in collaboration with a teacher who split the 10 students in the class in two equal groups, one participating under the traditional DBR approach and the other under the systems engineering-guided DBR approach. The teacher was familiarized with educational robotics and the traditional and systems engineering approaches to DBR through a summer professional development program.

(ii) Iterative design interventions for the systems engineering-guided DBR group [9]

A lesson on robot assembly was implemented as a part of engineering experience for students. Next, the assembled robot was employed in a math lesson on ratios. The objective of the experience was to familiarize students with basic engineering knowledge and skills and impart an initial understanding of vocabulary terms. The overall aim of the study was to inculcate interest and awareness among students for STEM disciplines through similar experiences. Table 4.3 explains the

TABLE 4.3 Iterative improvements in a middle-school robotics lesson using PDCA cycle [9]

	Iteration 1
Plan	Students were required to identify and sort out different pieces of the robotics kit and store them in different categories.
Do	While the teacher and a researcher guided the students through the activity, students were unable to complete the activity.
Check	The two reasons that contributed to students' inability to complete the activity were lack of introductory remarks and proper instruction sheets.
Act	For the next lesson it was determined that for a successful completion of the activity, the teacher needs to state learning objectives clearly and must arrange appropriate instructional material in advance.
	Iteration 2
Plan	Based on the recommendations from the "act" phase of iteration 1, the teacher took corrective actions by clarifying the learning objectives and providing instructional material in advance. During this lesson, the students were required to continue the same activity.
Do	Under the cognitive apprenticeship model, the researcher helped students to understand the functionalities of different components of the LEGO robot kit, interconnections between different parts, and various subassemblies. It was noticed that three out of five students were unable to complete subassemblies.
Check	The factors that most affected students' inability to complete subassemblies resulted from unsuitable instructional sheets that were printed black and white, instead of color, that may have confused the students about various components and lack of student attention during the explanation of various LEGO parts by the researcher.

Act	For the following lesson, it was recommended to arrange appropriate instructional sheets in color.
	Iteration 3
Plan	In this lesson, the students were required to complete the sorting activity and start building the robot.
Do	Despite the teacher and researcher guidance during the activity, the students could not accomplish the learning objective as desired.
Check	The lack of students' accomplishment was caused due to improper and inaccurate instruction sheets and time limitations.
Act	To improve the outcomes for the next lesson, the teacher and researcher collaboratively listed corrective action of reviewing the instructional sheet for its accuracy.
	Iteration 4
Plan	For this lesson, the students had to work toward the completion of robot assemblies.
Do	The teacher and researcher guided the students to build two robots. The students faced challenges in placing the LEGO brick in the appropriate location, making stable structures, selecting and placing gears, etc. The researcher showcased the correct assembly.
Check	The challenges in robot assembly resulted from students' lack of prior experience with robots and insufficient number of experienced adults to help the students during assembly.
Act	To cater to student needs on robot assembly, it was recommended to provide a sufficient number of instructional sheets and additional support to the students for the next lesson.
	Iteration 5—Math lesson
Plan	For this lesson, the teacher planned to illustrate the concepts of ratio and proportion while the students were required to observe and note the distance traveled by the robot for different gear ratios, use measuring tape and ultrasonic sensor for distance measurement, and compare the distance traveled for robots with different gear ratios.
Do	While the teacher guided students for the activity, they could not finish it in the given time. Additionally, the students struggled in finding correct gears and changing them and found a difference between activity sheets and robot behaviors.
Check	The reasons for incompletion of the desired plan largely stemmed from the teacher's lack of prior experience in doing similar activities and misjudgment of time needed to complete it. Moreover, there was an absence of mock practice of the activity and the activity sheets prior to the planned implementation, all of which caused challenges for the students in completing the gear activity.
Act	The teacher should plan for sufficient time and materials for the success of future lessons.

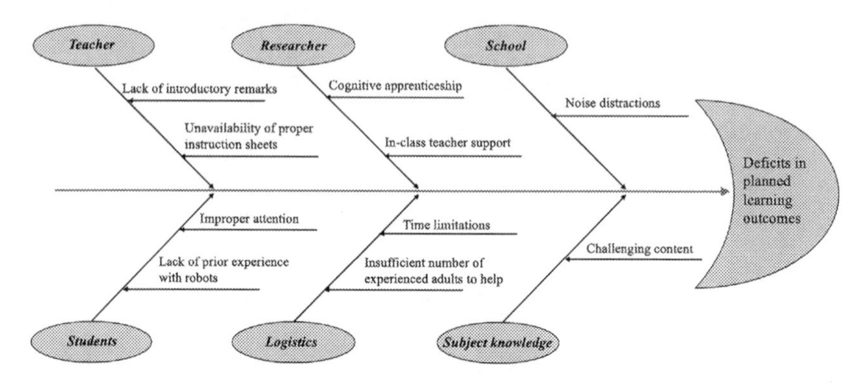

FIGURE 4.5 A cause-and-effect diagram (adapted from Rahman and Kapila [9]).

enhancement in the design of the lesson using the systems engineering approach through four iterative cycles for robot assembly and one iteration for a math lesson and Figure 4.5 illustrates the corresponding cause-and-effect diagram.

(iii) Outcomes of the study [9]

The engineering experience of assembling the robots introduced the students to different vocabulary terms such as control, gear, sensor, power, etc. All the students who participated in the robotics activities performed better and were more engaged in the presence of the DBR approach, irrespective of belonging to the traditional *versus* the systems engineering group. Yet, overall student outcomes for the group under the systems engineering approach were better in comparison to the ones in the traditional approach. For instance, participants in the systems engineering group made less mistakes in the activity and gained a better understanding of robot configuration than the students in the group under the traditional approach. Through student artifacts such as questionnaires and activity sheets, it was noticed that the students from the systems engineering group had more understanding about engineering vocabulary and accuracy in solutions than the ones in the traditional group. Finally, it was noted that the students from the systems engineering group demonstrated better learning outcomes as evidenced by their average score of 100% in a quiz related to ratios and proportions conducted after the robotics-based math lesson, as opposed to the students in the traditional group that achieved the average score of 60%. Successful updates to the design of this lesson can also be employed in other settings using similar systems engineering approaches.

(iv) Evolving principles [9]

Based on the iterative improvements and outcomes of implementing the systems engineering approach within a DBR framework, some evolving principles can be extrapolated and implemented in wider settings. They are delineated below.

1. Application of system engineering-based DBR iterations requires formal training of teachers followed by classroom support to effectively execute the methodology in real contexts. For example, the teacher who participated in this PDCA lesson improvement had prior experience of a summer professional development program and had an exposure to robotics. This enabled broader adoption of DBR principles and a readiness to participate in lesson design improvement.
2. The age level and grade appropriateness may require special considerations to incorporate DBR in K-12 settings. Teachers play a crucial role in suggesting those refinements to researchers. For example, the lack of introductory remarks and appropriate worksheets was identified earlier and the corresponding refinements mentioned above were critical in implementing effective lessons.
3. Even though robotics enables kinesthetic learning through its tangible platform, special considerations such as teacher guidance, well-paced and age-appropriate instructional methodologies, etc., must be employed to ensure that students get high-quality learning experiences. For instance, the use of scaffolded and colorful instructional sheets allowed the students to better understand and sort robotics components, something they were unable to do in the absence of appropriate instructional sheets.
4. Implementing robotic design and robotics-based math lessons can expose students to new knowledge such as basic engineering concepts, skills, and vocabulary, and may enhance their interest, awareness, and engagement in STEM, especially when such lessons are iteratively improved using the systems engineering approach. For instance, the outcomes of the study revealed a greater increase in engineering vocabulary and content knowledge of the students who belonged to the systems engineering group as opposed to the traditional group.

5. Implementation challenges of design-based research

Although DBR has been viewed as a method that can contribute to improving educational research, it is prone to certain challenges that may constrain its ability to produce scalable and sustainable solutions.

Design-based research methods necessitate the collection of data at each step of implementation, from multiple sources, and in varied formats, to gain a thorough understanding of the mechanics and impact of the intervention [4, 36]. This data may include observation videos, student interviews, student work artifacts, observation notes of researchers, etc. [5, 36], all of which together constitute the "thick descriptive datasets" [4] collected over the years of research with multiple participants. It is challenging to allocate adequate resources, in terms of personnel, time, effort, and finance, to analyze this data [36]. It is difficult and time-consuming to build a consensus around findings since each individual's interpretation of data may be colored by their personal views [37]. The DBR process is "over-methologized" since it uses only a small proportion of data to inform the findings and discards the rest without any use [37]. Overall, such a process of collecting, analyzing, and storing the data can be exhausting and thus poses an arduous challenge to any research team. The problems arising from excessive data may be remedied by deploying well-planned processes and instruments that adequately encapsulate the evolution of design intervention, capture its desired results, and obviate the need for storing and analyzing unnecessary data. Utilizing systems engineering tools and practices in tandem with the traditional method of DBR is one such approach that can produce favorable outcomes through focused data collection. The use of PDCA cycle delineates the efficacy of iterations over the traditional approach despite minimal data collection under rapid timelines.

A feature that distinguishes DBR from other education research methods is its mandate for collaboration between researchers and practitioners in developing, implementing, evaluating, and iterating the design of an intervention over a sustained duration. While this strategy enables researchers to gain an authentic overview of the context in which the research unfolds, it can be a formidable task to preserve these partnerships [4, 37]. The research formulated under the DBR framework frequently entails an intense endeavor lasting for one or more years, posing a difficulty for teachers, with on-going classroom and school obligations, to commit and contribute to multiple recursive cycles of refinement over long time durations [4, 37]. In fact, practitioners may often opt to enact a design *versus* to serve through the complete development cycle of an intervention [5]. Moreover, the collaborators on a DBR partnership may have divergent interests and motivations, for example, researchers may be keen to pursue an investigation informed by a theoretically-guided problem or solution strategy *versus* teachers may be intent on a practice-inspired exploration [37]. Thus, the work of cultivating a balanced and respectful partnership must be planned and calibrated, well in advance and with care, to prevent any undesired perturbation to the recursive cycles and the emerging results. Collaborations for the research scenarios of Section 4 were well-conceptualized to include motivated STEM teachers who themselves valued the opportunity to integrate robotics in their classrooms and actively participated in the professional development program to prepare for the same. Their prior teaching experience and enthusiasm supported smooth implementation of multiple cycles of DBR interventions. Nevertheless, there were several instances of time constraints and urgent requirements related to school curriculum, including test preparation mandates, that somewhat limited the achievement of desired outcomes.

Since grounded theories and propositions are indispensable to DBR [4], the lack of sufficient specificity of foundational underpinnings can cause failure in adequately illuminating the mechanics and impact of a design intervention [37]. In fact, such setbacks may be erroneously ascribed to a deficient implementation of the intervention *versus* to the unsound theoretical framing [37].

This absence of standards poses the dual challenges of deciding for how long should the research be continued and when does it become evident to be futile and thus fit for discarding [5, 37]. Finally, according to [4], DBR must not become "oversimplified," instead it should undergo continuous development to increase the rigor required for producing practical knowledge. For example, the three DBR scenarios of Section 4 resulted in positive research outcomes due to the presence of strong foundational theories, clear research vision, ongoing feedback processes, well-defined collaboration mandates, and high fidelity of implementation.

DBR interventions unfold over long durations during which researchers and practitioners collaborate to make a multitude of distinct choices [4]. Each such choice is accompanied with a number of factors that interact with and impact one another. In such a situation, it can be arduous to establish the causality of results and reliably tease out the factors that may be the accurate source of emerging principles and theories [4]. Moreover, as mentioned earlier, the absence of standards causes a relentless pursuit and production of new emergent theories without a sensible end in sight [37]. This situation makes it difficult to reliably isolate and link existing theories to new findings [4]. To overcome this challenge, systems engineering tools and methods can be applied as in Subsection 4.c. Such an approach can systematize the process of ascribing accurate causes for observed phenomena, and thus save time and resources.

Situating DBR in local contexts is the principal reason for the credibility of its findings. While progress in a local context is valuable for a specific design intervention under investigation, it may not be adequate for the design's adoption beyond those specific situations [8]. Instead, it is desirable that the emergent theories and principles of DBR be applicable to and benefit broader educational settings [8]. Moreover, even analogous settings can change over time. Thus, the designed educational practices ought to be relevant to, sustainable in, and generalizable for contexts beyond the one under investigation [8]. It may be particularly challenging to achieve a balance between two aspects of DBR, namely, suggesting design requirements to meet the practical usability requirements *versus* refining an educational theory to meet the goal of theory generation. Hence, DBR faces a risk in developing globally usable theories and principles due to its explicit focus on refining locally valuable designs. For example, the evolving principles from the three scenarios of Section 4 can be replicated in comparable technology-integration situations, yet their application is limited to similar learning contexts and depends upon the availability of resources.

The presence of researchers in classrooms supports authentic data collection and allows them to gain an in-depth understanding of real-world context. Yet, it can impact the research findings by influencing student and teacher behavior [5]. Specifically, being under observation may alter the nominal behavior of participants, leading to data collection that fails to represent the true situation and produces faulty research findings [5]. Moreover, researchers' involvement in the entire design cycle may make it difficult for them to serve as unbiased observers. In fact, they may intervene when participants struggle with something or their observations may be colored by their personal viewpoints [8]. Thus, the roles and responsibilities of the researchers in the implementation stage must be clearly delineated and controlled to prevent unintended contamination. For instance, the three DBR explorations of Section 4 included multiple yet focused methods to collect observational data through notes, recordings, and validated instruments [30, 38], to prevent individual biases and to obtain the most reliable outcomes.

6. Conclusion

Design-based research has been suggested as a methodology for bridging the gap between research theories inspired from laboratory investigations *versus* practical implementations [1, 2]. As researchers and practitioners collaborate to implement their designed intervention in a real-world context, the resulting outcomes are examined to serve as corrective refinements for the next iteration [1, 2]. Such a recursive process has the potential to enhance the intervention while generating effective theories and practices. Prior research has demonstrated that DBR produces desirable outcomes in the form of implementable theories and relevant practical knowledge [4, 5, 15]. This understanding from local contexts can be translated into other similar settings, adding value to a larger educational intervention [12].

Traditionally, the process of DBR applies recurring steps of analysis, design, enactment, and redesign [3–5] along with mixed methods research for data collection and analysis [5]. The typical DBR framework can be augmented with systems engineering tools using the PDCA paradigm of CIP cycles [33] to enhance the quality of iterations, as illustrated by the robot assembly lesson in Subsection 4.c. Using DBR to develop robotics-enabled STEM lessons can guide in ascertaining reliable approaches in promoting student motivation and can help in exploring the role of various learning theories such as cognitive apprenticeship, situated cognition, problem-based learning, and inquiry-based learning, among others [9].

Even though DBR can be a useful tool in developing theories and improving the design of an intervention, it also entails certain challenges including—collecting and analyzing large amounts of data [36], developing long-term partnerships [4, 37], pinpointing causality [4], and making generalizations [8]. Yet, DBR has been beneficial in advancing theories and practical knowledge, primarily in the fields of curriculum development, media and technology in education, instructional design, and teacher professional development [6, 7]. Thus, DBR explorations must be continued in validating educational products, processes, programs, or policies [7] in the times ahead.

7. Key takeaways

- The DBR paradigm emerged in response to the gaps identified by scholars between lab-based research and real-world implementation challenges [1, 2]. In addition to narrowing the credibility gap, DBR can aid in advancing theories of learning and teaching by generating more "usable knowledge" [12] and being more "socially responsible" [13].
- Some of the key attributes of DBR include: implementation in authentic settings in collaboration with practitioners [4, 5, 17], iterative nature [3–5], simultaneous focus on enhancing both theory and practice [4, 17], and use of mixed research methods [5].
- DBR contributes to both practice and theory. Specifically, its practical contributions include improvements in researcher-educator partnerships and increased relevance of research [4, 5, 15]. Moreover, its theoretical contributions give rise to three types of theories: domain theories, design frameworks, and design methodologies [16].
- Traditional DBR process [19] begins with identifying a relevant problem and a potential solution. Next, the prospective solution is field-tested in collaboration with practitioners and other participants [18]. Then, evaluation of the intervention is undertaken through comprehensive data collection and analysis after which necessary revisions to the design of the intervention are made [18]. Finally, researchers reflect on various aspects of the intervention to ascertain their impact and make amendments as necessary [18].
- In a systems engineering approach, DBR is treated as a CIP cycle that follows the PDCA paradigm [33] to improve the quality of iterations. The process is further strengthened by the use of Pareto principle, or the 80:20 rule [35], to find the most critical factors leading to the outcomes and the use of cause-and-effect diagrams [33] to pinpoint the exact causes behind the outcomes.
- While DBR can help in reaching a number of favorable outcomes, it can also pose certain drawbacks, for example, collecting and analyzing large amounts of data [36], developing long-term partnerships [4, 37], lacking a standardized process [4, 37], identifying causality [4], making generalizations [8], and clarifying different roles of participants [5], among others.

References

1. Collins, A., Toward a design science of education, in *New Directions in Educational Technology*, E. Scanlon and T. O'Shea, Editors. 1992, Berlin, Germany: Springer-Verlag. p. 15–22.
2. Brown, A.L., Design experiments: Theoretical and methodological challenges in creating complex interventions in classroom settings. *Journal of the Learning Sciences*, 1992.2(2): p. 141–178.

3. van den Akker, J., Gravemeijer, K., McKenney, S., and Nieveen, N., Introducing educational design research, in *Educational Design Research*, J. van den Akker, et al., Editors. 2006, London: Routledge. p. 1–8.
4. The Design-Based Research Collective, Design-based research: An emerging paradigm for educational inquiry. *Educational Researcher*, 2003.32(1): p. 5–8.
5. Wang, F. and Hannafin, M.J., Design-based research and technology-enhanced learning environments. *Educational Technology Research and Development*, 2005.53(4): p. 5–23.
6. van den Akker, J., Principles and methods of development research, in *Design Approaches and Tools in Education and Training*, J. van den Akker, et al., Editors. 1999, Dordrecht: Springer. p. 1–14.
7. McKenney, S. and Reeves, T.C., Contributions to theory and practice: Concepts and examples, in *Conducting Educational Design Research*. 2018, New York, NY: Routledge. p. 30–60.
8. Barab, S. and Squire, K., Design-based research: Putting a stake in the ground. *The Journal of the Learning Sciences*, 2004.13(1): p. 1–14.
9. Rahman, S.M. and Kapila, V., A systems approach to analyzing design-based research in robotics-focused middle school STEM lessons through cognitive apprenticeship, in *Proceedings of ASEE Annual Conference and Exposition*. 2017, Columbus, OH; Available from: https://peer.asee.org/27527.
10. Levin, J. and Donnell, A., What to do about educational research's credibility gaps? *Issues in Education*, 1999.5(2): p. 177–229.
11. National Research Council, Can research serve the needs of education? in *Improving Student Learning: A Strategic Plan for Education Research and Its Utilization*. 1999, Washington, DC: The National Academies Press. p. 9–20.
12. Lagemann, E.C., *An Elusive Science: The Troubling History of Education Research*. 2001, Chicago, IL: University of Chicago Press.
13. Reeves, T.C., Herrington, J., and Oliver, R., Design research: A socially responsible approach to instructional technology research in higher education. *Journal of Computing in Higher Education*, 2005.16(2): p. 96.
14. Moorhead, M., Listman, J.B., and Kapila, V., A robotics-focused instructional framework for design-based research in middle school classrooms, in *Proceedings of ASEE Annual Conference and Exposition*. 2015, Seattle, WA; Available from: https://peer.asee.org/23444.
15. McKenney, S. and Reeves, T.C., About educational design research, in *Conducting Educational Design Research*. 2018, New York, NY: Routledge. p. 7–30.
16. Edelson, D., Design research: What we learn when we engage in design. *The Journal of the Learning Sciences*, 2002.11: p. 105–121.
17. Anderson, T. and Shattuck, J., Design-based research: A decade of progress in education research? *Educational Researcher*, 2012.41(1): p. 16–25.
18. Scott, E.E., Wenderoth, M.P., and Doherty, J.H., Design-based research: A methodology to extend and enrich biology education research. *CBE—Life Sciences Education*, 2020.19(2): p. es11.1–es11.12.
19. McKenney, S. and Reeves, T.C., Core processes, in *Conducting Educational Design Research*. 2018, New York, NY: Routledge. p. 83–180.
20. Papavlasopoulou, S., Giannakos, M.N., and Jaccheri, L., Exploring children's learning experience in constructionism-based coding activities through design-based research. *Computers in Human Behavior*, 2019.99: p. 415–427.

21. Kopcha, T.J., McGregor, J., Shin, S., Qian, Y., Choi, J., Hill, R., Mativo, J., and Choi, I., Developing an integrative STEM curriculum for robotics education through educational design research. *Journal of Formative Design in Learning*, 2017.1(1): p. 31–44.
22. Rahman, S.M., Krishnan, V.J., and Kapila, V., Fundamental: Optimizing a teacher professional development program for teaching STEM with robotics through design-based research, in *Proceedings of ASEE Annual Conference and Exposition*. 2018, Salt Lake City, UT; Available from: https://peer.asee.org/30551.
23. Ackermann, E., Piaget's constructivism, Papert's constructionism: What's the difference? in *Proceedings of Constructivism: Uses and Perspectives in Education*. 2001, Geneva, Switzerland: Research Center in Education. p. 85–94; Available from: https://learning.media.mit.edu/content/publications/EA.Piaget%20_%20Papert.pdf.
24. Savery, J.R. and Duffy, T.M., Problem based learning: An instructional model and its constructivist framework. *Educational Technology*, 1995.35(5): p. 31–38.
25. Collins, A., Cognitive apprenticeship and instructional technology, in *Educational Values and Cognitive Instruction: Implications for Reform*, L. Idol and B.F. Jones, Editors. 1991, Mahwah, NJ: Lawrence Erlbaum Associates, Inc. p. 121–138.
26. Collins, A., Brown, J.S., and Newman, S.E., Cognitive apprenticeship: Teaching the craft of reading, writing and mathematics. *Thinking: The Journal of Philosophy for Children*, 1988.8(1): p. 2–10.
27. Cognition Technology Group at Vanderbilt, Anchored instruction and its relationship to situated cognition. *Educational Researcher*, 1990.19(6): p. 2–10.
28. Eguchi, A., Educational robotics theories and practice: Tips for how to do it right, in *Robots in K-12 Education: A New Technology for Learning*, B.S. Barker, Editor. 2012, Hershey, PA: IGI Global. p. 1–30.
29. Ferdig, R.E., Assessing technologies for teaching and learning: Understanding the importance of technological pedagogical content knowledge. *British Journal of Educational Technology*, 2006.37(5): p. 749–760.
30. Moorhead, M., Elliott, C.H., and Listman, J.B., Professional development through situated learning techniques adapted with design-based research, in *Proceedings of ASEE Annual Conference and Exposition*. 2016, New Orleans, LA; Available from: https://peer.asee.org/25967.
31. Lave, J. and Wenger, E., *Situated Learning: Legitimate Peripheral Participation*. 1991, New York, NY: Cambridge University Press.
32. Gerard, L.F., Varma, K., Corliss, S.B., and Linn, M.C., Professional development for technology-enhanced inquiry science. *Review of Educational Research*, 2011.81(3): p. 408–448.
33. Kuznetsova, E.S., Continuous improvement process as a way to increase effectiveness, in *Proceedings of IEEE Russian-Korean International Symposium on Science and Technology*. 2004, Tomsk, Russia. p. 244–246.
34. Li, Y., Li, X., and Li, J., Exploring the underlying mechanism of PDCA cycle to improve teaching quality: A motivation theory perspective, in *Proceedings of IEEE Portland International Center for Management of Engineering and Technology Conference: Infrastructure and Service Integration*. 2014, Kanazawa, Japan. p. 2693–2698.
35. Gittens, M., Kim, Y., and Godwin, D. The vital few versus the trivial many: Examining the Pareto principle for software, in *Proceedings of IEEE Annual International Computer Software and Applications Conference*. 2005, Edinburgh, Scotland. p. 179–185.
36. Collins, A., Joseph, D., and Bielaczyc, K., Design research: Theoretical and methodological issues. *The Journal of the Learning Sciences*, 2004.13(1): p. 15–42.

37. Dede, C., If design-based research is the answer, what is the question? A commentary on Collins, Joseph, and Bielaczyc; diSessa and Cobb; and Fishman, Marx, Blumenthal, Krajcik, and Soloway in the JLS special issue on design-based research. *The Journal of the Learning Sciences*, 2004.13(1): p. 105–114.
38. You, H.S., Chacko, S.M., and Kapila, V., Examining the effectiveness of a professional development program: Integration of educational robotics into science and mathematics curricula. *Journal of Science Education and Technology*, 2021.30(4): p. 567–581.

5

EFFECTIVE PROFESSIONAL DEVELOPMENT FOR ROBOTICS-FOCUSED LEARNING ENVIRONMENTS

1. Introduction

Teacher professional development is a critical factor in expanding the understanding, abilities, mindsets, and practices of teachers and it serves as an influential contributor to enriching the quality of teaching and learning in schools [1]. With the proliferation of various technology tools in educational settings, the need to upskill teachers, raise their confidence, and help them use the same is gaining momentum [2]. Lacking a formal exposure to contemporary technology tools [3], facing challenges in staying updated with the ongoing innovations [2], and having minimal guidance on the hands-on use of technology tools [3] all contribute to teachers' apprehension in incorporating technology for instruction. Moreover, a lack of resources and experience with ineffectual design of professional development programs [4] may cause teachers to employ technology tools merely for administrative purposes as opposed to using them as pedagogical supports [3]. Thus, it is necessary to offer teachers professional development opportunities that are embedded with features rendering them effective in creating a sustained change [5, 6] and that are flexibly structured [5] to meet their intended goals and requirements. To successfully infuse technology tools such as robotics in instruction, teachers must participate in hands-on active learning experiences that can instill confidence in them to not only understand the technology but also use it to enhance pedagogy.

Effective professional development experiences [1]: develop teachers' knowledge, skills, and mindsets; support them to implement the newly acquired knowledge, skills, and mindsets to improve their instruction; and engender advancements in teaching methods that enhance student learning outcomes. Creating and facilitating such high-quality effective professional development experiences, especially

DOI: 10.4324/b23177-7

ones that meet the needs of teachers employing modern tools in their classrooms, requires the presence of some core features that include: form, ample duration, collective participation, engaging and new content, active learning, and coherent design [6]. These hands-on active learning experiences can boost teachers' confidence and help them in infusing technology to enhance instruction.

Designing and implementing robotics-based professional development programs, in particular, require a specific focus on understanding teachers' current levels of preparedness and concerns about managing classrooms with robots [7]. In addition to meeting the criteria for effective professional development [6], it is necessary to address specific teacher concerns, give them ample hands-on opportunities for enriching their knowledge, skills, and attitudes about integrating robots in their instruction, and situate their learning [8] in real-world contexts. In seeking to design authentic and sustained teacher learning opportunities, this chapter first describes different professional development models and relevant characteristics that make it effective. Then, following a brief literature review on robotics-based learning for teachers, two case studies on professional development for robotics-enhanced environments are detailed [3, 9]. The first study [3] employs the situated learning theory [8] with activities that focus on problem solving using real-world situations and involve collaborative group work and reflection. The second study [9] applies the social capital theory [10] to build relationships between teachers, as they co-design lessons, leading to improved skills in lesson planning. For the two case studies, the chapter showcases design, implementation, and outcomes of the professional development experiences. The characteristics and design of the aforementioned studies can be tailored to empower teachers as they participate in professional development programs in varied contexts.

2. Teacher professional development

Teacher professional development is "a complex array of interrelated learning opportunities" delivered through varied methods and activities that enrich the understanding, abilities, and mindsets; instructional practices; and socio-emotional repertoire of teachers [1]. In recent years, policymakers and educators have initiated systemic reforms to amend the traditional ways of teaching and learning. With teachers being situated at the forefront of deploying these new initiatives successfully, it is imperative that they are adequately equipped with disciplinary content knowledge and pedagogical abilities to communicate and develop advanced skills among students. Since teacher proficiency in delivering instruction is directly linked to improved student learning [1], it plays a critical role in enacting progressive education reforms [6].

As technology permeates our everyday life, including teaching and learning, there is a growing demand to prepare and support teachers in integrating technology into their classrooms [3]. Teachers' aversion to employ technology in their lessons is caused by their lack of experience with advanced technology tools and it poses a significant hindrance in creating technology-enhanced learning environments [3]. The rapid evolution in the hardware and software of various advanced technologies can pose a deterrence for teachers to keep abreast with them [2]. Furthermore, the provision of none to minimal support for practical use of technology in professional development programs can cause teachers to ignore the promise of technology for instructional tasks in favor of administrative purposes [3]. Thus, teachers must be afforded training and practice opportunities to meaningfully integrate advanced technology solutions for teaching and learning. In fact, research has shown that teachers who participate in professional development programs that explicitly address technology integration gain knowledge, skills, and confidence in implementing technology-based lessons [2].

To become effective as a teacher of science or math discipline, one must attain a strong foundational understanding of the disciplinary concepts, gain fluency in the pedagogical content knowledge, and hone skills to situate lessons in realistic settings [11]. Next, to cater to the ever-evolving needs of students by using sophisticated instructional methodologies; adopting hands-on and inquiry-based learning; and promoting 21st-century skills of critical thinking, communication, collaboration, and creativity; it is paramount to prepare and support teachers through professional development [4, 12]. Moreover, learning standards, such as the Next Generation Science Standards (NGSS) [13], have shifted the focus toward the development of higher-order thinking skills among students. Thus, there is an urgent demand to significantly alter conventional teaching practices by equipping teachers to effectively implement the new standards. It is suggested that a meaningful enactment of these standards necessitates that teacher professional development programs incorporate hands-on activities, with personal relevance to students, that are also responsive to the standards [9]. To promote the adoption of appropriate instructional methodologies, professional development programs must focus on the concepts taught via the curricular standards, corresponding pre-requisite knowledge, and the expected needs of classroom and students as perceived by the teachers [9, 14]. Overall, in the pursuit of improving the quality of hands-on science, technology, engineering, and mathematics (STEM) education in schools, teachers must be offered effective and timely professional development opportunities with a focus on technology integration, STEM content knowledge, instructional methodologies, and standards alignment.

2.a. Models of professional development [5]

Professional development refers to activities, programs, and interventions that engage teachers in enhancing their knowledge, abilities, beliefs, and practices during the course of their careers [15, 16]. A professional development program can be structured in multiple ways depending on its intended purposes and contexts [5]. Such experiences can expand the content understanding, skills, attitudes, and instructional practices of teachers as well as enrich their personal and socio-emotional repertoire [1].

Kennedy [5, 17] developed a framework to examine eight models of professional development: training, deficit, cascade, award-bearing, standards-based, coaching/mentoring, community of practice, and collaborative professional inquiry. These eight models can be grouped into three overarching categories namely, transmissive, malleable, and transformative, based on the fundamental purposes of professional development and in the increasing order of teacher capacity for professional autonomy. A brief description of the three categories and eight models is provided below.

Category 1: Transmissive professional development programs focus on training teachers to implement education reforms. This category consists of the *training, deficit*, and *cascade* models of professional development.

- The *training model* is a popular type of professional development that supports skills-based learning through which teachers get opportunities to upskill and showcase their abilities. Delivered by experts who determine the agenda, training serves as an effective means to introduce new knowledge to the teachers. In alignment with standards-based professional development, training avails the teachers of opportunities to showcase specified abilities designated in the standards under consideration. Despite being an effective means to transfer knowledge, training may fail to illustrate practical uses of new learning. With the training being commonly located off-site, it may be divorced from the classroom context and lack connection to authentic situations. Finally, the standardization of professional development under the training model may obscure the need for teachers to be proactive participants in designing learning experiences based on their individual needs.
- The *deficit model* targets perceptions of deficits in teacher performance for remedy. By failing to consider collective responsibility for teacher performance, this model often falls short. Specifically, in attributing the root cause of poor teacher performance solely to the teacher, it neglects the ill effects of organizational and management practices on teacher performance.
- The *cascade model* entails training events for engaging individual teachers who in turn are expected to disseminate information to their colleagues. In a resource-limited environment, the cascade model can be a valuable asset. Despite its promise to help in the transfer of knowledge and skills, this model does not address teaching values and attitudes. Moreover, when the direct participants of this model disseminate their learning, they may discard relevant strategies or settings for learning, all of which may limit the deep processing of new information by their colleagues.

Category 2: Malleable professional development programs can be structured as either transmissive or transformative, based on their intended implementation [17].

This category consists of the *award-bearing, standards-based, coaching/mentoring*, and *community of practice* models of professional development.

- The *award-bearing model* entails teachers finishing an award-bearing program of study, usually from a post-secondary institution. Even as the model is characterized by quality assurance, its emphasis on classroom practice *versus* values and mindset development may limit its potential for deep learning.
- The *standards-based model* envisions a teacher education system that can empirically inform the relationship between teacher effectiveness and student outcomes. The model provides a "common language" for teachers to discuss ideas and practices. Informed by the behaviorist learning theory, it seeks to link rewards in response to improvement in teacher competence and, thus, is not collaborative. Moreover, the model lacks focus regarding the purpose of teaching and limits teachers in having alternative conceptions.
- The *coaching/mentoring model* is based on a one-on-one relationship between a dyad of teachers discussing ideas, problems, possibilities, and beliefs. While coaching supports improvement in the abilities of one peer in the dyad by the other, mentoring entails counseling and support for a novice member by an experienced member of the dyad. The ability to embody high-quality interpersonal communication skills by the participants is essential for the success of this model.
- The *community of practice model* fosters teacher learning through interactions within a community that comprises more than two people. Collaborative learning through this model can lead to the creation of knowledge. Based on the behaviors of participants in a wider community, learning under this model can be either active or passive.

Category 3: Transformative professional development programs emphasize supporting teachers to refine policy and practice, and consist of the *collaborative professional inquiry* model.

- The *collaborative professional inquiry model* involves teachers as researchers to engage in enhancing the quality of teaching and learning. Such a model can lead to greater teacher participation and makes learning more relevant and active.

It is important to note that the above list of the types of professional development models is not exhaustive. The models can be modified or integrated to serve the vision and requirements of teachers.

3. Designing for effective professional development

Teacher professional development is one of the primary sources of teacher learning, skill-building, and personal growth and, thus, needs to be effective and meaningful to create a demonstrable and sustained change. To be successful, a professional development experience must [1]:

- Enhance teachers' understanding and abilities, alter their attitudes and mindsets, or both.
- Enable teachers to implement the gained knowledge, abilities, attitudes, and mindsets to enhance instructional content, pedagogy, or both.
- Lead to change in instructional methodology that improves student learning outcomes.

3.a. Structural features [6]

A high-quality professional development program has three features that define its structure or design namely, *form*, *duration*, and *collective participation*.

Form: It describes the structure of the program by defining whether it has a traditional or reform orientation. Traditional programs consist of workshops or conferences, which are situated outside real classroom contexts, do not provide sufficient opportunities to make meaningful changes to teacher practices, and thus are usually not well-received by teachers [6, 18]. Reform programs such as study groups, mentorship, individual research projects, etc., are situated in school settings and can realistically inform classroom practices. By being more content focused and by embedding active learning opportunities, reform programs entail a tighter alignment to the needs and contexts of teacher work [6, 18]. The traditional form of programs can also be made more effective by incorporating core features (see Subsection 3.b) [18].

Duration: Professional development experiences that endure over a long time can be more effective in two specific ways: (1) they allocate teachers adequate time to explore the content and address misconceptions and (2) they allow teachers to attempt new practices, gather feedback, interact with other participants, and reflect [6]. In our experience, a range of 20–160 contact hours of professional development is effective and a longer duration is more likely to have positive outcomes for teachers.

Collective participation: Grouping teachers from similar grades, subjects, or schools can lead to a number of advantages such as opportunities to discuss skills, concepts, practices, curriculum, and assessment requirements and specific student needs based on their backgrounds [6, 18]. A professional development program directed at a single school can make learning more enduring by cultivating professional camaraderie. In such an intervention, the group can discuss common concerns, best practices, knowledge, etc. This can further help in establishing organizational learning and promoting individual teacher growth [6, 18].

3.b. Core features [6]

Studies have found some common attributes, called 'core features,' that are essential for designing effective and successful professional development experiences. Irrespective of the *form*, these features can be a driver to promote effective teacher learning [1]. Drawing from the comprehensive research by Garet et al. [6], these features are summarized in the following paragraphs.

Content: This refers to the subject knowledge, skills, and practices that the participants will gain through the course of a professional development program. The program content may include disciplinary knowledge that the teachers are supposed to deliver and pedagogical approaches that they may employ in their practice. That is, the professional development content may comprise of disciplinary subject matter, instructional methods (e.g., classroom management, lesson planning), and/or pedagogical content knowledge, which combines pedagogical knowledge with specific content knowledge to make the learning of concepts seamless. The content for an effective professional development also builds teacher capacity to employ specific curriculum material or teaching strategies and it incorporates activities that support teachers in enhancing skills and conceptual understanding of students. Furthermore, the content builds teacher understanding of how students learn by addressing their cognitive levels, misconceptions, and subject-specific strategies. Professional development experiences that focus on improving the subject matter expertise of teachers and relying on how students learn can help in improving classroom instruction, eventually leading to positive student learning outcomes.

Active learning: This fundamental attribute focuses on imparting active learning opportunities to teachers via engagement in content, discussions, planning, and practice. Active professional development experiences include participants observing master teachers, being observed teaching in their own classrooms by mentors or master teachers, gathering feedback about their practices, and participating in reflective discussions. These active experiences afford the participants opportunities to integrate new ideas in their classroom contexts such as student expectations and previous experiences, grade levels, school situations, etc. In an active professional development program, the participants examine student work to learn about prior conceptions and reasoning abilities of students. This can allow them to design lessons with appropriate difficulty levels and cater to multiple needs of students. Finally, the participants can deepen their understanding via teacher-led presentations, discussions, and written work.

Coherence: This refers to the alignment between individual activities within a program. A professional development activity is more likely to have a positive outcome if it is a coherent component of a larger endeavor. Following are the three ways to ascertain whether the activities are coherent or not. Coherent activities build on the previous set of activities and are followed by more advanced work; are aligned with the national, state, and local standards and assessment; and promote communication among teachers to share methods, challenges, and reflections on similar issues.

Overall, the key features that help in designing high-quality and effective professional development experiences are—*duration*, *collective participation*, *content*, *active learning*, and *coherence*. The *form* of activities, traditional and reform, have comparable outcomes if sustained for similar durations or if they embody the core features and, thus, is not identified as a primary characteristic.

4. Literature review on teacher professional development for robotics-based learning

A comprehensive review [19] on research projects related to educational robotics distilled 28 studies that focused on teacher professional development. The review identified that potential gaps in disciplinary and instructional expertise of teachers can hamper the effective integration of robotics for learning and, thus, emphasized the need for constructionist teacher professional development experiences [20]. The synthesis of studies showcased that educational robotics-focused constructionist professional development programs can: improve teachers' instructional methodology and, in turn, enhance students' academic performance; develop the self-efficacy and knowledge of teachers in introducing robotics in their classrooms; and render opportunities for teachers to design, improve, and contextualize robotics-focused curriculum pertaining to their classrooms [19]. Following is a brief description of three teacher professional development programs that focused on integrating robotics for teaching and learning.

Northeastern University, Tech-Boston (a part of the Boston Public Schools), and Tufts University's Center for Engineering Education Outreach designed and conducted a two-week professional development program for 12 middle school teachers [21]. While the primary focus of the program was on introducing a robotics unit to teachers using a hands-on, experiential learning approach, it additionally afforded them opportunities for peer learning, discussions, and teaching a group of four to six students. The first week of the program entailed the teachers learning from guest presenters and engaging in hands-on explorations concerning a 10-lesson robotics unit. In the second week, during the first half of each day, the teachers taught the lessons from the first week to a group of students, and during the second half of each day they attended additional workshop sessions. The teachers had several opportunities to reflect and discuss their takeaways with peers and the project staff. The results of the professional development assessment indicate that the teachers had enhanced their content knowledge of engineering fundamentals and their confidence in teaching the same to their students. Moreover, the results reveal that the teachers valued hands-on engagement and appreciated learning from the experiences of other teachers. To bring a significant change in teachers' pedagogical styles, the researchers noted the need to consider strategies that create hands-on inquiry-based explorations for teachers so that they can impart similar learning experiences to their students.

Beyond Blackboards, an after-school engineering program for middle school students was conceptualized and implemented by research faculty and mentors from the University of Texas at Austin [22]. The program was founded on a three-way integrated approach that engaged students, teachers, and caregivers to enhance their awareness of STEM and its related future pathways. One of the three key participating groups included teachers who were recruited for a week-long professional development institute and subsequent academic year activities. The institute aimed to prepare the teachers to support students in purposeful engineering design activities and in exploring future STEM pathways. Following the five-step model of *hands-on exploration, interactive discussion, exploratory laboratories, open-ended design challenges*, and *reflection*, the teachers engaged in engineering design and project-based learning activities through robotics technology. Additionally, the teachers received exposure on instruction for programming, design-based teaching and learning, and details of the curriculum through collaborative experiences. The focus group interactions with the teachers revealed that they perceived positive changes in themselves regarding instruction and envisioned myriad skill-development possibilities through robotics for their students.

The University of Texas at San Antonio designed and implemented an interdisciplinary robotics-focused teacher professional development workshop by leveraging Technology Acceptance Model 3 (TAM3) for elementary and secondary school teachers [23]. In the context of incorporating robotics as a technology tool for classroom use, TAM3 suggests that a teacher will employ robotics when: it is *perceived as useful* via factors such as improved student outcomes, alignment with standards, etc., and it is *easy to use*, that is, when teachers perceive higher self-efficacy, enjoyment, and usability, etc. Ten teachers, of whom eight had prior experience with using robotics in after-school programs, participated in the workshop. Primed by TAM3, the workshop considered three primary elements that included: avenues for aligning standards to robotics activities, utilizing project-based learning to apply an inquiry-based approach to address student requirements, and supporting teachers' pedagogical skill improvement. The post-workshop survey results indicate that the teachers were able to make explicit connections of the learned robotics activities with classroom-based teaching content. Moreover, the new teachers expressed an improved interest and greater ease in using robotics as a learning tool for their classroom. The interdisciplinary approach of this study can aid in engaging teachers and, thus, their students in learning through robotics.

5. Designing a robotics-based professional development program using situated learning

Despite understanding the promise of educational robotics to promote STEM learning, teachers feel unprepared and remain averse to implement it in their classrooms [2]. Robotics-based professional development programs that fail to intentionally account for the burden of managing a classroom with robots, integrating disciplinary content with robotics activities effectively, understanding the interplay between instructional strategies and robotics technology, etc., may derail classroom enactment and further exacerbate the situation [7]. Thus, to help alleviate teachers' lack of preparedness, it is essential that robotics-focused professional development programs adequately attend to understanding teacher requirements; enriching their knowledge, abilities, mindsets, and attitudes about robotics integration; and meeting the criteria of effectiveness. In this spirit, this section describes the objectives, theoretical concepts, designs, and outcomes of a robotics-based professional development program that was founded on the concept of situated learning [8].

The robotics-based teacher professional development program was designed and conducted over a three-year period to prepare over 40 middle school teachers to meaningfully deploy robotics-enhanced STEM lessons in their classrooms. The program aimed to prepare teachers by enhancing their technological pedagogical content knowledge (TPACK) [24], guiding them to develop engaging standards-aligned science and math lessons using robotics kits, modeling effective lesson examples using robotics as a technological tool, and supporting lesson implementation in their classrooms. Informed by the features of effective professional development [1, 6, 18], various professional development models [5], and examples from existing literature, this program was founded on the theories of situated learning [8], constructionism [2, 25], and cognitive apprenticeship [2, 26, 27]. Design-based research [28] was employed to refine the program and the classroom lesson activities introduced during the program.

The professional development program of [3] was a pilot implementation with four teachers consisting of two pairs of science and math teachers from two middle schools. The program was implemented during summer over three weeks via 15 sessions (eight hours per session) and was led by facilitators consisting of a team of engineering and education faculty, researchers, and graduate students. The initial design for the pilot program consisted of three modules—introduction to LEGO Mindstorms EV3 robotics kits, collaborative group activities to experience and refine standards-aligned lessons, and review and understanding of education research literature. A systematic program assessment revealed [2] a statistically significant impact on teachers' TPACK self-efficacy [29, 30] and content knowledge for robotics. The program's theoretical underpinnings, structure, and outcomes are delineated below.

5.a. Theoretical concepts

This professional development program employed design-based research to iteratively refine its design and improve lessons, as well as TPACK, situated learning, constructionism, and cognitive apprenticeship to facilitate effective teacher learning. This subsection describes the theories and their aligned implementation for this program.

Design-based research (DBR) process employs iterative improvements to the learning design to develop usable learning theories and artifacts [28]. DBR entails collaborative implementation of prospective solutions in authentic settings to help enhance both theory and practice [28]. Applying DBR in robotics-enhanced learning environments can help in identifying the most reliable and viable approaches to increase student motivation and engagement [31]. In the program of [3], DBR was used to make two distinct iterative refinements: enhancement of robotics-based STEM learning plans and improvement in the design of professional development programs.

Technological pedagogical content knowledge (TPACK) [24] draws on the interactions and relationships between three distinct domains of knowledge, namely, technology, content, and pedagogy, to deepen teaching and learning in a technology-enabled educational setting. It provides an effective framework for teachers to conceive productive ways to incorporate technology in teaching and learning. To achieve such an outcome, the TPACK framework characterizes seven different types of knowledge that need to be effectively imbibed, including: content knowledge, pedagogical knowledge, technological knowledge, pedagogical content knowledge, technological content knowledge, technological pedagogical knowledge, and technological pedagogical content knowledge. When a technology is integrated in instruction ineffectively, it can lead to a superfluous use of technology that can cause the lesson to deviate from its goal. This professional development program aimed to improve teachers' TPACK to prepare and motivate them for integrating robotics effectively in STEM teaching.

Situated learning [2, 8] entails learners participating in authentic activities in the context or situation in which they typically occur. The learning opportunities are embedded with activities that require problem-solving, replicate real-world situations, and involve collaborative group work and reflection. The use of a situated learning framework in a professional development program focused on robotics-based lessons can aid in developing teachers' instructional practices and in analyzing their beliefs about teaching with robots. Thus, this professional development program was designed in the context of situated learning where a group of researchers and teachers collaboratively: identified and honed conceptual knowledge and skills required to use educational robotics as a pedagogical tool, developed and tested lessons that use robots in real-world contexts, and prototyped and examined robotics creations for specific learning situations. DBR was employed to investigate the effectiveness of using the construct of situated learning for professional development.

Constructionism [2, 25] is a learning theory where students create their own knowledge by building or making tangible artifacts. This form of learning is active in nature, advocates making connections between varied concepts, and is facilitated by the teacher through coaching rather than didactic instruction. This approach can be applied in designing and enacting robotics-enabled lessons that afford learners opportunities to perform hands-on problem-solving and receive feedback, by observing the performance of their robotics device, to improve their learning. Based on the theory of constructionism, this professional development program was embedded with multiple hands-on activities, for example, designing a base robot, incorporating different sensors and understanding their operation, programming robotics devices, experiencing and refining robotics-based sample lessons, developing new lesson plans, etc.

Cognitive apprenticeship [2, 26, 27] enables a novice learner to acquire cognitive and metacognitive abilities to problem-solve in real-world contexts under the guidance of a domain expert. Six strategies can facilitate this type of learning: *modeling* by illustrating the thinking of an expert, *coaching* by aiding learners during activities, *scaffolding* by embedding support to enable a learner to perform a certain task, *articulating* by supporting learners in clarifying their thinking approach, *reflecting* by assisting learners to examine their performance, and *exploring* by fostering learner autonomy in gaining conceptual understanding. Cognitive apprenticeship was used during the professional development program by engaging teachers in engineering practices under the expert guidance of robotics engineers and education researchers.

5.b. Structure of the professional development program [3]

Successful professional development programs that seek to promote technology-enhanced instruction and learning are characterized by hands-on experiential learning, constructivist approach to learning, and enduring assistance to participants for one or more years [32]. The program of [3] was developed by building on the features of effective professional development along with strong theoretical foundations. To best enable the teachers to gain comfort with robotics and to give them sufficient time, the program consisted of three modules spanning over a period of three weeks (120 hours).

- Module 1 engaged teachers in hands-on, active learning via an introduction to the LEGO Mindstorms EV3 robotics kits. The focus was for the teachers to understand different components of the kit, functionalities of various sensors, programming, etc.
- Module 2 entailed collaborative group activities to practice and refine standards-aligned lessons. Here the focus was for participants to model the designed lessons, examine their intended standards alignment, and understand potential impediments that students might encounter.
- Module 3 encouraged teachers to review and understand learning content including education research literature. Specifically, as teachers were expected to participate in iteratively refining the professional development program under the DBR framework, it was pertinent to expose them to

relevant educational research constructs. Hence, they were introduced to the underlying research and participated in discussions on the topics of DBR, TPACK, situated learning, cognitive apprenticeship, intrinsic and extrinsic motivation, problem and project-based learning, etc.

The following is a breakdown of weekly activities of these modules during the implementation of the program. The activities were refined through DBR.

Week 1: The teachers participated in hands-on activities to acquaint themselves with the LEGO Mindstorms EV3 robotics kit and discussed research on TPACK. To enable situated learning, they were introduced to learning units on modeling, energy, and expressions and equations through PowerPoint presentations and related hands-on robotics activities. Robot-building instructions were provided to each team for self-paced learning. The teachers completed all the activities except for introducing some minor iterations. It was observed that they found it difficult and time-consuming to follow build instructions and these were subsequently amended. Teacher-led collaborations gave rise to a new base robot design that the teachers deemed more suitable for students.

Week 2: The teachers read and discussed research readings on DBR. They were introduced to additional lessons on ratios and proportions, analyzing and interpreting data, statistics, and functions. They participated in related hands-on activities and developed aligned assessment material for their classrooms. With increased time to interact with their robot, the teachers were able to understand its capabilities and recognize it as a pedagogical tool on their own. The teachers collectively contributed to a revised design for a lesson where the LEGO Mindstorms EV3 software was doing all the work, such as creating scaled plots, while minimizing student involvement. The revised lesson design allowed teachers to hone their robotics integration skills which fostered opportunities for enhanced student involvement in learning through robots.

Week 3: The teachers were introduced to articles on intrinsic and extrinsic motivation, project-based learning, and cognitive apprenticeship. They were encouraged to develop new robotics-based lesson plans, activities, and assessments using two approaches: a content-driven approach where they identify a topic that students find challenging to comprehend and a technocentric approach where they use a robot sensor they weren't introduced to yet. The teachers collaboratively thought of challenging topics such as addition and subtraction of integers and developed lesson plans. A comprehensive instructional guide was developed after it was observed that the teachers found it challenging to navigate the programming interface.

Finally, different professional development models (see Subsection 2.a) were considered and integrated to develop and execute this program in the most effective, engaging, and relevant way. The program began with the *training* model under which the facilitators delivered instruction and the teachers experienced lessons and activities as students. As teachers gained a level of comfort in

(a)

(b)

FIGURES 5.1 (a)–(b) Teachers engaged in performing robotics-based learning activities.

performing robotics activities and as DBR-based iterations were enacted, scaffolds were progressively eliminated and the program became more active. Thus, the program evolved to resemble the *collaborative professional inquiry* model and the *mentoring and coaching* model as it sought to advance teacher knowledge, skills, and confidence. Moreover, as the overall program design embedded discussions around theoretical concepts and lesson design, a *community of practice* emerged that could sustain beyond the summer professional development program. Details on the step-by-step improvement to the professional development program can be found in Chapter 4 on DBR. See also Figures 5.1 (a)–(b) illustrating teachers engaged in performing robotics-based learning activities.

5.c. Outcomes [2]

It is important to evaluate the effectiveness of a professional development program upon its completion. Moreover, it is essential to identify features that lead to the success of a program so that educators can use suitable tools and methodologies for future programs and continue to refine the aspects that were inadequate [1]. As mentioned in Section 3, an effective professional development experience enables teachers to build and enact new knowledge, abilities, mindsets, and attitudes as well as leads to positive student outcomes [1]. Effectiveness of professional development programs can be measured using surveys, interview data, classroom observations, student learning outcomes, etc. [1].

The pilot program of [3] was developed keeping the criteria of effectiveness at the forefront. It was scheduled for an adequate time duration, focused on hands-on experiential learning via robotics kits and lesson development activities, involved collaborative group work for lesson refinement and understanding theories, and was coherent with prior content knowledge and standards alignment. Having conducted the pilot professional development program with four teachers, it was scaled and conducted with 43 teachers over the next two years [33]. The scaled-up program's effectiveness was measured using three instruments—TPACK self-efficacy survey, content knowledge test, and reflection questions and interviews. The TPACK self-efficacy survey [2] (see sample items in Section 3 of Appendix A) consisted of all seven TPACK constructs and was adapted to include the dimensions of confidence, motivation, outcome expectancy, and apprehensiveness. The content knowledge technical quiz [2, 3] was administered as a pre- and post-test instrument to gauge teacher knowledge gains about STEM and robotics topics. Finally, the reflection and interview prompts obtained teachers' perceptions toward program engagement, their learning, facilitator expertise, and their own enthusiasm.

As evidenced in [33], pre- and post-test analysis of the TPACK self-efficacy surveys produced a statistically significant improvement in the dimensions of perceived confidence and outcome expectancy. However, even as the teachers experienced a drop in the apprehension dimension, the results are not statistically significant. The teachers also demonstrated a statistically significant growth in their content knowledge, moving from 46.13% mastery prior to *versus* 57.6% mastery after the professional development program, yielding approximately 24.86% improvement. The reflection response data indicated that the teachers viewed the program positively concerning engagement, learning outcomes, satisfaction, facilitator expertise, successfulness, and attitude. The reflection and interview data revealed five major themes of: teachers learning new skills and knowledge, increased teacher motivation and/or confidence, improved student learning, enhanced student motivation and interest, and opportunity to collaborate with other teachers. Following the completion of the professional development program, through classroom observations, it was noted that teachers were efficient in loading programs and identifying programming difficulties. Overall, data analysis revealed that the professional development program met its goals of improving teachers' TPACK, science and math content knowledge, and improvement in independently developing and implementing lessons using LEGO Mindstorms EV3.

6. Creating a professional development program using the social capital theory

Professional development goes beyond attending formal, structured programs and includes learning in informal settings with other teachers and through classroom interactions with students [1]. Embedding collaborative learning opportunities can make a professional development program more relevant and effective for teachers [5]. The effort of [9] created a professional development program under the theory of social capital [10] to generate collaborative learning and foster resource sharing among the program participants in terms of information, feedback, and knowledge to support the members of the social group. A three-phase summer professional development program was organized with science and math teachers to co-develop a science curriculum by integrating NGSS, 5E instructional model [34], and robotics to address the STEM learning needs of middle school students. The program of [9] used Bourdieu's theory of social capital [10] and the construct of critical constructivism [35] to create collaborative learning environments. The curriculum was co-developed through multiple phases: facilitator-led professional development; NGSS curriculum co-development during teacher professional development, and teachers as facilitators training. The program's theoretical concepts, structure, and outcomes are described below.

6.a. Theoretical concepts

The program of [9] was grounded in the sociocultural theoretical framework of social capital that was utilized to co-design a curriculum for robotics-infused STEM lessons and critical constructivist perspective was used to describe the balance of power in the mentor-protégé relationship. The program employed a novel NGSS-plus-5E lesson design model for hands-on robotics-based explorations.

Social capital [10]: It is the aggregate generated by the investment of members of a network to build trust and to create ways for mutual recognition and acknowledgment. Investment can consist of resources such as information, support for other individuals, intellectual capital, and influence. Social capital can make all other types of capitals (economic, political, etc.) and resources accessible to the network group. When applied in the context of teacher development, mentor–protégé relationships (i.e., teacher–teacher or facilitator–teacher relationships), in the purview of social capital theory, can make specialized knowledge more accessible to protégés. For the professional development program under consideration, teachers and facilitators shared human capital that enabled social learning, ultimately leading to the development of intended instructional practices. It is important to select a diverse group of participants for such a collaboration to facilitate better knowledge sharing.

Critical constructivism: It "involves a complex interrelationship between teaching and learning and knowledge production and research" [35]. Critical constructivists recognize the need for consciously addressing educational theories that lead to a deeper reflection on how they can be improved and implemented [35]. For this program, the purpose of critical constructivism was to enable the facilitators and teachers to build their own learning and consciousness through the process of co-developing the curriculum.

NGSS-plus-5E model: Many students often experience a chasm between what they learn at school *versus* what they experience in the real world [36]. Additionally, they do not find classroom learning personally relevant, which makes them lose intrinsic and extrinsic motivation, especially in STEM subjects [37]. NGSS were created as an effort to respond to this challenge. Specifically, NGSS were developed after rigorous research to support students in developing deeper and holistic science learning by engaging in explanatory models, scientific investigations, and observations [9, 14]. They are significantly different from traditional science teaching approaches and, thus, require professional development programs for a smooth transition for teachers. Moreover, the 5E instructional model (see Chapter 9) helps teachers organize their lessons to make learning more active and to facilitate conceptual change. Incorporation of robotics in STEM classrooms, along with NGSS and 5E model, can make learning more effective, engaging, and personally meaningful for students. This professional development program introduced the teachers to the NGSS-plus-5E lesson planning model and utilized the social capital developed through varied interactions in improving the design of lessons.

6.b. Structure of the professional development program

Traditional professional development programs often engage teachers in learning and replicating a pre-specified set of content and activities as opposed to adapting various materials and creating their own curriculum. Involving teachers in developing and testing their own lessons can engage them in designing curriculum units that are contextualized, for student interest, classroom setting, etc., and endow them with a sense of ownership. Thus, this program focused on co-design of NGSS-plus-5E robotics curriculum and it was conducted in three different phases: facilitator-led professional development; NGSS curriculum co-development during teacher professional development; and teachers as facilitators training.

Phase 1—Facilitator-led professional development: This phase involved an engineering faculty member and an education researcher mentoring program *facilitators*, namely, two engineering graduate students and two engineering postdoctoral researchers. Having first provided the facilitators with self-study material on the theoretical frameworks for the program, after two weeks, the mentors led a one-day workshop on the NGSS and 5E instructional model. As a follow-up to this workshop, the facilitators reviewed previously designed robotics-enhanced science and math lessons to ascertain their NGSS gaps. Next, under the guidance of the mentors, over a week-long period, each facilitator sought to enhance a lesson for closing the NGSS alignment gap by using an NGSS-plus-5E lesson template provided to them. Finally, the facilitators presented four NGSS-plus-5E lessons to their peers and the mentors to receive feedback and then performed lesson modifications (see Figure 5.2).

Phase 2—Teacher professional development for co-design of NGSS-plus-5E lessons: For this phase, six *participants* were selected from a group of teachers who had previously collaborated in a robotics-based professional development program. These participants included three science and three math teachers who attended the program for three weeks (five days for eight hours a day). In addition to

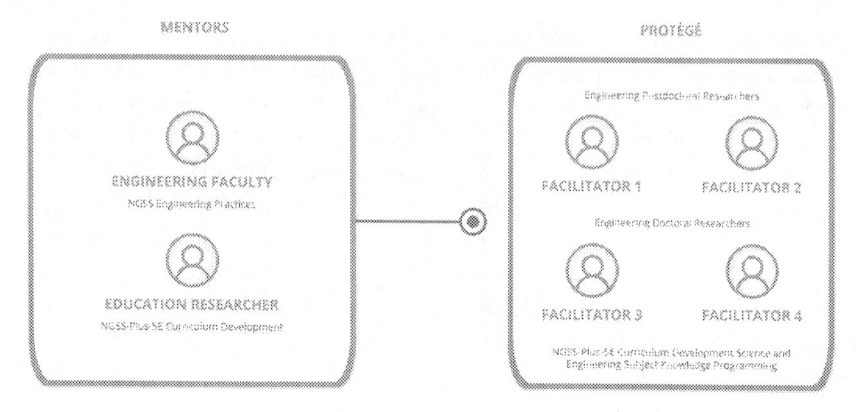

FIGURE 5.2 Mentor-protégé relationships in Phase 1 (adapted from [9]).

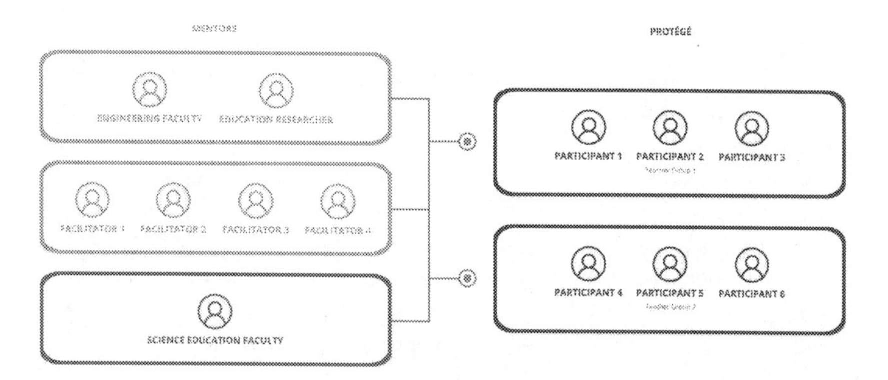

FIGURE 5.3 Mentor-protégé relationships in Phase 2 (adapted from [9]).

the activities outlined below, each of the weeks consisted of LEGO robotics and programming lessons to refresh and update participant knowledge (see Figure 5.3).

- **Week 1:** The first day of the program focused on an NGSS-plus-5E module led by the education mentor. The remaining four days were used to demonstrate to the participants the four lessons developed during Phase 1, one lesson a day. Each day's lesson demonstration was followed by discussions on the adequacy of its NGSS alignment and suggested improvements were made to the lesson plan.
- **Week 2 and 3**: In week 2, the six participants were divided into two groups. Each group was tasked to improve a previously designed and used robotics-focused lesson plan to align it with the NGSS standards. After engaging for two days in the lesson improvement process, on the third day, each participant

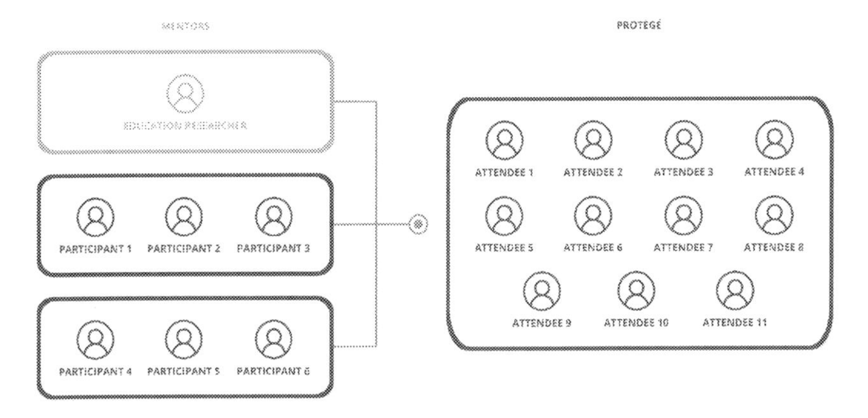

FIGURE 5.4 Mentor-protégé relationships in Phase 3 (adapted from [9]).

group received feedback on their plans from the facilitators, mentors, and their peers. Next, on the fourth day, each participant group modified their lesson plans in response to the feedback, resulting in two NGSS-plus-5E lessons. A similar process was employed for week 3, resulting in two additional lesson plans.

Phase 3—Teachers as facilitator training: For this phase, 11 teachers were selected as *attendees*. They attended the one-day NGSS-plus-5E workshop conducted on the first day of first week in phase 2. They also attended the fifth days of second and third weeks when the six participants from phase 2 demonstrated their finalized NGSS-plus-5E lessons. These 11 attendees acted as protégés and gave feedback on the lessons to their mentors, that is, the six participants. This phase helped in building relationships amongst both sets of teachers (see Figure 5.4).

6.c. Outcomes

During the course of the social capital-based professional development program, data collection items included: daily reflections of the four facilitators and the six participants, extensive field notes taken by the four facilitators, and pre- and post-surveys of the teachers (six participants and 11 attendees) regarding their NGSS knowledge. The data was analyzed during the program to improve its quality for the on-going cycle and summatively after the program to improve its future offerings. The analysis of data led to the identification of the following three major themes: teacher resistance to NGSS, challenges of NGSS-plus-5E implementation, and science and math identity.

During and after the one-day introductory workshop in the second phase, teacher resistance to learn about the vision, developmental process, and structure of NGSS persisted. A primary driver for this was teachers' lack of clarity about the standards, which was remedied only through their active engagement in lesson development. Thus, the post-program data showed that teachers gained familiarity with the NGSS and 5E instructional model. The teachers initially found it challenging to effectively address the three dimensions of NGSS in their robotics-based lessons. Even as the social capital generated through discussions allowed teachers to surmount their initial hesitation, the one-day introductory workshop was determined to be inadequate. Implementation of lessons under the 5E model within the typical 45-minute class period was determined to be too cumbersome, thus these lessons were split into smaller chunks to make the lessons more manageable. Math teachers were seen losing interest in developing NGSS-plus-5E lessons, however, with facilitator mentorship, they were able to learn how they could identify and include math components in these lessons using the standards. Similarly, the facilitators, who were engineering graduate and postdoctoral researchers, were able to develop a greater appreciation for teachers' work.

The program was designed using the principles of effective professional development programs (see Section 3) as it: sustained for a sufficient duration of time; promoted active learning through discussions, activities, and presentations; focused on new content sharing; stimulated collective participation; and was imbued with coherence in its organization and implementation. Overall, the application of the social capital theory resulted in strong ties between the teachers and facilitators, contributed to designing new lessons, and improved teachers' skills in lesson planning.

7. Challenges in planning effective professional development programs and incorporating their lessons

Although the research-guided features of effective professional development [1, 6, 18] can be adopted to design meaningful and relevant experiences for professional development programs, studies [4] have identified the following barriers that can limit the success of a program. While some of these impediments to the success of the program may occur during its planning phase, others may transpire when integrating its lessons in the school schedule.

Ascertaining professional development needs and outcomes: Many professional development programs do not perform needs assessment to establish teacher requirements and they lack a collective understanding of superior teaching. With the daily operation of schools being their primary concern and absent any formal training, many administrators lack proficiency in recognizing, understanding, and organizing high-quality professional development for their teachers [4, 38]. To respond to this obstacle, which induces a misaligned vision, the social capital-based professional development program (see Section 6) focuses on collaboratively designing the curriculum and gives opportunities to teachers to voice their requirements and ideas [9]. Furthermore, it is crucial to track the outcomes of the professional development program implementations to surmount the barriers and build on the successes for future designs. Absence of this step may occur due to a lack of time and resources. Results from outcomes analysis can add to useful information about the tools and methods that can meaningfully engage teachers and have the potential to improve larger practice [4, 38]. The professional development programs of Sections 5 and 6 employed DBR and used both formative and summative evaluation techniques to gather outcomes with the intent to improve the programs in future iterations [9, 33]. This made the programs more robust for both—the teachers who were enrolled in it and the ones who may join in its future iterations.

Ineffective professional development design: Unfortunately, due to time constraints, teacher unavailability, and lack of resources, some professional development programs fail to satisfy the critical factors identified in Section 3 that influence the effectiveness of a professional development program. To meet the aforementioned logistical constraints, often the structure, content, or activities of a program need to be amended, which may cause deficiencies in its high quality and high fidelity implementation. To ensure program implementation with effectiveness and fidelity, it is essential to be cognizant of the possibility of logistical constraints and mitigate their adverse effects with strategies that obviate the need to compromise the structure, coherence, and time duration of the program. For example, the professional development programs of Sections 5 and 6 were conducted during three weeks of summer, recruited and secured participant commitment in late spring, explicitly communicated roles of and expectations from participant to them and their schools, acquired all necessary robotics materials prior to summer, and collaborated with a group of middle school teachers to conceive, create, and test a pilot curriculum for the program, among others. Nonetheless, even when the design of a professional development program satisfies the requirements for its effectiveness, it may still suffer from deficiencies in its implementation with high quality and high fidelity. These deficits may arise when a professional development program is devoid of a logical approach, which accounts for the complexity of a teacher's work with competing priorities, and suffers from inadequate capacity of professional development providers or participants [4, 38]. For example, in the two-week professional development of [21], the teachers found it challenging to connect with and manage students that they had not known previously. This hindered teachers from concentrating on the robotics-based lesson as they had to respond to interference created by these students. Moreover, an efficacy analysis of professional development based on the social capital theory showed that a one-day workshop on the NGSS-plus-5E model was insufficient to prepare the teachers [9]. The teacher challenge of connecting with and managing new students can be ameliorated by having the teachers recruit students from their own classrooms [39]. Moreover, the problem of insufficient time to adequately familiarize with fundamental concepts of a professional development model can be addressed by performing a trade-off between time and effort allocated to the acquisition of principles *versus* the engagement in practice.

Lack of time and resources: Due to their crowded schedule at school, teachers are often not allocated adequate time to enact the new knowledge and skills acquired from participating in a professional development program. Moreover, in the absence of easy access to resources such as curriculum material, technology infrastructure, learning artifacts, etc., teachers frequently have to draw on their own time and finances to procure such materials. This produces an inequitable divide between students since some teachers may have the resources to acquire the materials while others may not. This barrier is more pervasive in schools located in high-poverty, resource-starved neighborhoods [4]. As one illustration of time limitation, in one of the follow-up in-class observations, wherein DBR was being implemented using a system's approach (see Chapter 4), it was seen that the teacher had limited time to implement the planned lesson that led to failure in achieving the desired goals [31]. To overcome such hurdles, a professional development program must systematically cultivate relationships with school communities, secure their commitments for allocating adequate resources and time, and encourage them to co-create a shared vision of learner success. Moreover, aligning intended lessons with prevailing learning standards, as opposed to enacting them as standalone learning activities, can be a catalyst for teachers in resourcefully utilizing their classroom time. To facilitate collective learning and execution of learning from robotics-based professional development program of Section 5, the research team partnered with schools by inviting two teachers from each school so that they could learn collaboratively, share resources, and rely on each other for solidarity.

8. Conclusion

Teacher professional development plays a key role in improving the quality of learning in schools. With increased focus on technology integration and standards-alignment, this need has been amplified. To design high-quality and meaningful professional development experiences, certain core features must be incorporated including, sufficient duration, collaborative participation, introduction to engaging and new content, active learning, and a coherent plan. Inefficient design can lead to failure in meeting the intended professional development goals and teacher inability to effectively enact the new knowledge and skills. This chapter presents two methods of developing, refining, and executing effective professional development programs under the purview of integrating robotics in middle school STEM classrooms. One of the programs is founded on situated learning, giving hands-on opportunities to teachers to develop their TPACK. The other program is based on the theory of social capital wherein collaborative knowledge of teachers and facilitators results in social learning and, thus, improved lesson planning skills. The program designs can be adopted and implemented in similar settings that use emerging technologies to enhance student learning.

9. Key takeaways

- Teacher professional development is "a complex array of interrelated learning opportunities" delivered through a variety of methods and activities that contribute to teachers' knowledge and skills, instructional expertise, and personal growth [1].
- There are eight models of professional development according to Kennedy's framework: training, deficit, cascade, award-bearing, standards-based, coaching/mentoring, community of practice, and collaborative professional inquiry [5].
- The core features that make a professional development program effective are: sufficient duration, collective participation, engaging and new content, active learning experiences, and coherent design [6].
- As discussed in the case studies above, the use of situated learning [8] in robotics-based lessons [3] can develop teachers' practices and help in analyzing their beliefs about teaching with robots since such opportunities include activities that focus on problem-solving, use real-world situations, and involve collaborative group work and reflection. Additionally, the application of the social capital theory [10] can lead to stronger relationships between the teachers and facilitators, help in co-developing lessons, and improve teachers' skills in lesson planning [9].
- Some barriers that can limit the promise of a professional development program include the lack of time and resources, inadequately ascertaining professional development needs, ineffective design, and failure in determining professional development program outcomes [4, 38], among others.

References

1. Desimone, L.M., A primer on effective professional development. *Phi Delta Kappan*, 2011.92(6): p. 68–71.
2. You, H.S. and Kapila, V., Effectiveness of professional development: Integration of educational robotics into science and math curricula, in *Proceedings of ASEE Annual Conference and Exposition*. 2017, Columbus, OH; Available from: https://peer.asee.org/28207.
3. Moorhead, M., Elliott, C.H., Listman, J.B., Milne, C.E., and Kapila, V., Professional development through situated learning techniques adapted with design-based research, in *Proceedings of ASEE Annual Conference and Exposition*. 2016, New Orleans, LA; Available from: https://peer.asee.org/25967.
4. Darling-Hammond, L., Hyler, M.E., and Gardner, M., *Effective Teacher Professional Development*. 2017, Palo Alto, CA: Learning Policy Institute.

5. Kennedy, A., Models of continuing professional development: A framework for analysis. *Journal of In-service Education*, 2005.31(2): p. 235–250.
6. Garet, M.S., Porter, A.C., Desimone, L., Birman, B.F., and Yoon, K.S., What makes professional development effective? Results from a national sample of teachers. *American Educational Research Journal*, 2001.38(4): p. 915–945.
7. Krishnan, V.J., Rajguru, S.B., and Kapila, V., Analyzing successful teaching practices in middle school science and math classrooms when using robotics (Fundamental), in *Proceedings of ASEE Annual Conference and Exposition*. 2019, Tampa, FL; Available from: https://peer.asee.org/32092.
8. Anderson, J.R., Reder, L.M., and Simon, H.A., Situated learning and education. *Educational Researcher*, 1996.25(4): p. 5–11.
9. Ghosh, S., Krishnan, V.J., Rajguru, S.B., and Kapila, V., Middle school teacher professional development in creating a NGSS-plus-5E robotics curriculum (Fundamental), in *Proceedings of ASEE Annual Conference and Exposition*. 2019, Tamps, FL; Available from: https://peer.asee.org/33108.
10. Bourdieu, P., The forms of capital, in *Handbook of Theory and Research for the Sociology of Education*, J. Richardson, Editor. 1986, New York, NY: Greenwood Publishing Group. p. 241–258.
11. Krishnamoorthy, S.P., Rajguru, S.B., and Kapila, V., Fundamental: A teacher professional development program in engineering research with entrepreneurship and industry experiences, in *Proceedings of ASEE Annual Conference and Exposition*. 2018, Salt Lake City, UT; Available from: https://peer.asee.org/30547.
12. Dede, C., Comparing frameworks for 21st century skills, in *21st Century Skills: Rethinking How Students Learn*, J. Bellanca and R. Brandt, Editors. 2010, Bloomington, IN: Solution Tree Press. p. 51–76.
13. NGSS, *Next Generation Science Standards (NGSS): For States, By States*. 2013, Washington, DC: The National Academies Press; Available from: https://www.nextgen science.org/.
14. Reiser, B.J., What professional development strategies are needed for successful implementation of the next generation science standards, in *Proceedings of Invitational Research Symposium on Science Assessment*. 2013, Washington, DC. p. 1–22; Available from: http://www.ets.org/Media/Research/pdf/reiser.pdf.
15. Lessing, A. and De Witt, M., The value of continuous professional development: Teachers' perceptions. *South African Journal of Education*, 2007.27(1): p. 53–67.
16. Dilshad, M., Hussain, B., and Batool, H., Continuous professional development of teachers: A case of public universities in Pakistan. *Bulletin of Education and Research*, 2019.41(3): p. 119–130.
17. Kennedy, A., Understanding continuing professional development: The need for theory to impact on policy and practice. *Professional Development in Education*, 2014.40(5): p. 688–697.
18. Birman, B.F., Desimone, L., Porter, A.C., and Garet, M.S., Designing professional development that works. *Educational Leadership*, 2000.57(8): p. 28–33.
19. Anwar, S., Bascou, N.A., Menekse, M., and Kardgar, A., A systematic review of studies on educational robotics. *Journal of Pre-College Engineering Education Research (J-PEER)*, 2019.9(2): p. 2.
20. Alimisis, D. and Kynigos, C., Constructionism and robotics in education, in *Teacher Education on Robotic-enhanced Constructivist Pedagogical Methods*, D. Alimisis, Editor. 2009, Athens, Greece: School of Pedagogical and Technological Education (ASPETE). p. 11–26.

21. Hynes, M. and Dos Santos, A., Effective teacher professional development: Middle-school engineering content. *International Journal of Engineering Education*, 2007.23(1): p. 24.
22. Crawford, R.H., White, C.K., Muller, C.L., Petrosino, A.J., Talley, A.B., and Wood, K.L., Foundations and effectiveness of an after-school engineering program for middle school students, in *Proceedings of ASEE Annual Conference and Exposition*. 2012, San Antonio, TX; Available from: https://peer.asee.org/21404.
23. Zhou, H., Yuen, T.T., Popescu, C., Guillen, A., and Davis, D.G., Designing teacher professional development workshops for robotics integration across elementary and secondary school curriculum, in *Proceedings of IEEE International Conference on Learning and Teaching in Computing and Engineering*. 2015, Taipei, Taiwan. p. 215–216.
24. Mishra, P. and Koehler, M.J., Technological pedagogical content knowledge: A framework for teacher knowledge. *Teachers College Record*, June, 2006.108(6): p. 1017–1054.
25. Kafai, Y.B., Constructionism, in *The Cambridge Handbook of the Learning Sciences*, R.K. Sawyer, Editor. 2005, Cambridge, UK: Cambridge University Press. p. 35–46.
26. Collins, A., *Cognitive Apprenticeship and Instructional Technology*. 1988, Cambridge, MA: BBN Labs, Inc. 16.
27. Collins, A., Brown, J.S., and Newman, S.E., Cognitive apprenticeship: Teaching the craft of reading, writing and mathematics. *Thinking: The Journal of Philosophy for Children*, 1988.8(1): p. 2–10.
28. The Design-Based Research Collective, Design-based research: An emerging paradigm for educational inquiry. *Educational Researcher*, 2003.32(1): p. 5–8.
29. Schmidt, D.A., Baran, E., Thompson, A.D., Koehler, M.J., Mishra, P., and Shin, T., *Survey of Preservice Teachers' Knowledge of Teaching and Technology*. 2009; Available from: https://sciencetonic.de/media/015_digimedia/015_tpack/LIT_110_Schmidt_Baran_Mishra_Koehler_et_al_TPACK_Survey_2009.pdf.
30. Schmidt, D.A., Baran, E., Thompson, A.D., Mishra, P., Koehler, M.J., and Shin, T.S., Technological pedagogical content knowledge (TPACK) the development and validation of an assessment instrument for preservice teachers. *Journal of Research on Technology in Education*, 2009.42(2): p. 123–149.
31. Rahman, S.M. and Kapila, V., A systems approach to analyzing design-based research in robotics-focused middle school STEM lessons through cognitive apprenticeship, in *Proceedings of ASEE Annual Conference and Exposition*. 2017, Columbus, OH; Available from: https://peer.asee.org/27527.
32. Gerard, L.F., Varma, K., Corliss, S.B., and Linn, M.C., Professional development for technology-enhanced inquiry science. *Review of Educational Research*, 2011.81(3): p. 408–448.
33. You, H.S., Chacko, S.M., and Kapila, V., Examining the effectiveness of a professional development program integration of educational robotics into science and mathematics. *Journal of Science Education and Technology*, 2021.30(4): p. 567–581.
34. Bybee, R.W., Taylor, J.A., Gardner, A., Scotter, P.V., Powell, J.C., Westbrook, A., and Landes, N., *The BSCS 5E Instructional Model: Origins and Effectiveness*. 2016, Colorado Springs, CO: Office of Science Education, National Institutes of Health.
35. Steinberg, S.R., Critical constructivism, in *The SAGE Encyclopedia of Action Research*, D. Coghlan and M. Brydon-Miller, Editors. 2014, London: SAGE. p. 203–206.
36. Symonds, W.C., Schwartz, R., and Ferguson, R.F., Pathways to prosperity: Meeting the challenge of preparing young Americans for the 21st century, in *Pathways to Prosperity Project*. 2011, Cambridge, MA: Harvard University Graduate School of Education.

37. Brickhouse, N., Bringing in the outsiders: Reshaping the sciences of the future. *Journal of Curriculum Studies*, 1994.26(4): p. 401–416.
38. Tooley, M. and Connally, K., No panacea: Diagnosing what ails teacher professional development before reaching for remedies. *New America*, 2016. p. 1–41.
39. Mallik, A., Liu, D., and Kapila, V., Analyzing the outcomes of a robotics workshop on the self-efficacy, familiarity, and content knowledge of participants and examining their designs for end-of-year robotics contests. *Education and Information Technologies*. 2022; Available from: https://doi.org/10.1007/s10639-022-11400-1.

6

APPLYING TPACK TO DESIGN FOR ROBOTICS-ENHANCED LEARNING

1. Introduction

With the accelerating emphasis on engaging students in learning and preparing them for the workforce in science, technology, engineering, and mathematics (STEM) fields, educators have become increasingly receptive to incorporating educational technologies in classrooms. Educational technologies are recognized as possessing the promise of fostering student-centered, active learning environments [1]. Although technology-infused education can provide myriad benefits in learning and teaching, embracing it without a thoughtful consideration of curriculum content and student learning needs can produce a shallow *technocentric* integration of technology [2, 3]. Specifically, technocentric integration bestows significance primarily to a selected technology, including its promise and constraints, and only secondarily does it scrutinize how the technology may boost pedagogical and cognitive outcomes [4]. Since a technocentric approach does not foreground the technology integration process with an inquiry into how technology can best illuminate disciplinary content, it fails to adequately and explicitly address the matters of pedagogical content knowledge (PCK) [5].

Pedagogical content knowledge entails the ability to use "the most powerful analogies, illustrations, examples, explanations and demonstrations," to transform foundational concepts of a discipline into representations deemed accessible by learners [5]. However, in contrast to various disciplinary content that K-12 students are required to learn, many STEM concepts are abstract, lacking in tangible and accessible exemplars [4]. In response, the integration of technology in the classroom must be driven primarily by making STEM concepts more accessible, meaningful, and relevant and additionally advancing student STEM learning through deep and engaging active learning. In fact, the rise in the adoption of

DOI: 10.4324/b23177-8

technology to enable effective teaching and learning of STEM disciplines has broadened the construct of PCK to yield the framework of "technological, pedagogical, and content knowledge" (TPACK) [6, 7]. Under the TPACK framework, teachers aid their students to engage with and learn abstract and complex disciplinary content with technology serving as a tool. Specifically, the TPACK framework fosters effective uses of technology for STEM concepts, which are often devoid of concrete and accessible depictions, by rendering alternative knowledge representations that students find meaningful [7].

In this chapter, we illustrate that using robotics to design STEM learning experiences offers a promising approach that fosters active learning, makes abstract concepts more accessible, provides multiple modes of presenting content, and promotes collaboration [8]. When used in conjunction with the TPACK framework, teachers can effectively design robotics-based lessons to generate hands-on experiences that allow the learners to actively construct their knowledge [9]. In this chapter, we address the design of effective technology-integrated learning experiences by exploring the TPACK framework while synergistically embedding the criteria proposed by Earle [10] and Ferdig [11] for examining such integration. In addition to introducing the theoretical foundations of the framework and literature on TPACK-guided interventions, the chapter showcases several examples in which educational robotics plays the role of technology in TPACK formalism. Specifically, the chapter introduces examples of a professional development program that built teachers' robotics-related TPACK, and several robotics-infused STEM lessons grounded in the TPACK paradigm. Design and adoption of lessons, activities, and programs produced in this vein can aid teaching and learning to be informed by and infused with constructivist methods [12] and represent a departure from the mere technocentric aspects of technology-based learning.

2. Technological, pedagogical, and content knowledge

TPACK is a framework that characterizes the knowledge that educators require to effectively integrate technology in the service of disciplinary content with instructional pedagogy [7]. It refers to "an understanding of the representation of concepts using technologies; pedagogical techniques that use technologies in constructive ways to teach content; knowledge of what makes concepts difficult or easy to learn and how technology can help redress some of the problems that students face; knowledge of students' prior knowledge and theories of epistemology; and knowledge of how technologies can be used to build on existing knowledge to develop new epistemologies or strengthen old ones" [6]. Effecting synergistic interactions among the three TPACK domains can deliver on the promise of technology as a practical pedagogical tool that effectively renders novel representations of disciplinary knowledge deemed more acceptable than the conventional ones by students [4]. In fact, according to the proponents of

TPACK, the endeavor of technology infusion in the classroom obligates teachers to gain fluency not only in the three TPACK domains but also in their complex intersections to create meaningful learning experiences [6]. With an ineffective integration, the classroom will be subject to a naïve use of technology and result in the departure of lessons from their intended purposes.

2.a. Components of the TPACK framework

For its three foundational knowledge domains, TPACK characterizes various intersecting relationships that are purposeful and mutually-dependent, and that can be used to enrich technology-infused learning experiences. As seen in Figure 6.1, the interactions among the three knowledge domains generate four interrelated components, namely, pedagogical content knowledge (PCK), technological content knowledge (TCK), technological pedagogical knowledge (TPK), and technological pedagogical content knowledge (TPACK). Thus, the TPACK framework comprises seven knowledge components that are delineated below [3, 6, 7] along with their examples.

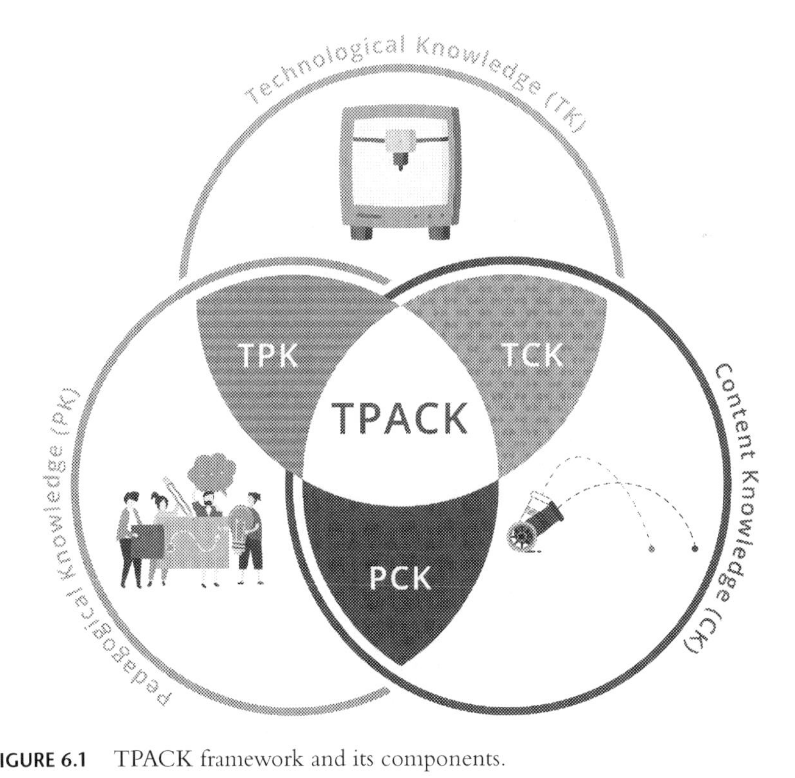

FIGURE 6.1 TPACK framework and its components.

Credit line: Image designed using assets from Flaticon.com and Freepik.com.

Content knowledge (CK) refers to disciplinary knowledge and the difference in the nature of knowledge for various content areas. It includes the understanding of the facts, models, norms, contexts, and routines of a discipline that are addressed in the learning environment. The lack of command of disciplinary content may induce inaccurate teaching and misconceptions among students. As one illustration of CK, prior to launching a unit on ratios and proportions, a math teacher should be conversant with the facts, rules, and procedures corresponding to the unit to circumvent any possible misrepresentation.

Pedagogical knowledge (PK) refers to the knowledge about the techniques and routines for effective teaching and learning and it requires a profound understanding of the classroom to flexibly apply theories of learning and cognitive science. It consists of the vision and purpose of a class, process to manage and assess it, capacity to plan and execute a lesson, and ability to foster student learning. Such knowledge is essential to design and deliver learning in a way that permits students to acquire, retain, and apply their learning. For example, in introducing the unit on ratios and proportions with embedded activities, the math teacher should account for the ideal group-size for the activity and design for effective teamwork.

Technology knowledge (TK) refers to the knowledge of various technological artifacts, including routines, techniques, skills, and tools to design, produce, and operate them. It includes knowledge of technologies that span from low-tech (e.g., tape measure and craft materials) to digital technologies (e.g. augmented and virtual reality headsets and mobile learning apps). The ability to adapt a technology effectively for learning requires cultivating a level of comfort with its myriad features and applications. For example, to introduce ratios and proportions with educational robotics in an experiential learning setting, the math teacher should become well-versed with the various components and features of the robots, such as construction pieces, chassis, drive train, sensors, motors, controller, and programming interface, among others.

Pedagogical content knowledge (PCK) refers to the knowledge of disciplinary content, viewed from the lens of teaching and learning process [5], and it tailors pedagogical techniques to enhance learner understanding of the content. It encompasses teaching strategies that invoke prior knowledge of students, address their misconceptions, and embed germane representations, analogies, and exemplars of concepts (e.g., a phenomenon described verbally, depicted pictorially, portrayed in a tabular form, or even enacted by students, all to personalize learning experiences for differently-abled learners). Introduced in Shulman's seminal work [5], PCK enriches the repertoire of teachers' understanding and skills to guarantee that their students can make sense of the content and remedy their prior misconceptions. For example, in the lesson on ratios and proportions, the math teacher may embed real-world examples to elicit students' perceptions about and their computational routines for ratios while making the learning relevant and personalized.

Technological content knowledge (TCK) refers to the knowledge concerning effective use of particular technologies to exemplify, understand, and practice material related to specific disciplinary content. It includes knowledge of myriad features and modes of technologies that can be used to enrich the understanding of facts, theories, and procedures from a particular discipline. Command of TCK domain allows teachers to utilize the affordances of technology tools in representing the content tangibly and with meaning, while also preventing its unfruitful use. For example, in the lesson on ratios and proportions, the math teacher may focus on the robot drive train built with gears so that the students are deeply engaged in hands-on learning for constructing their understanding of ratios and proportions. Educational robotics, like other technologies, is not a panacea for every pedagogical challenge related to abstract and complex disciplinary concepts. Indeed, if the attributes of an education robotics activity appear to address the intended content of a lesson only superficially, then, the teacher should apply her TCK to seek an alternative technology tool that better serves the learning goals of the content.

Technological pedagogical knowledge (TPK) refers to the knowledge of various technology tools and their affordances that can aid in specific instructional tasks. It is an understanding that allows teachers to resolve whether a technology can advance a chosen pedagogical strategy. Rather than conferring technology a central role in their lesson, teachers with such knowledge effectively align technology use with their pedagogy to deepen student learning. For example, returning to the ratios and proportions lesson, the math teacher must enact their TPK to examine how educational robotics can support effective teamwork and make the learning more connected to real life.

Technological pedagogical content knowledge (TPACK) refers to the knowledge required to integrate technology with the other core domains of content and pedagogy. It includes the knowledge of affordances (or limitations) of technologies that can improve (or impede) teaching of a disciplinary concept in a pedagogically sound (or unsuitable) manner. This interweaving of all the three components is the basis of engendering quality learning experiences that use technologies appropriately and in relevant contexts. For instance, in the ratios and proportions lesson, the math teacher may consider various features and functionalities of educational robotics that can foster a learning community among students, with freedom of expression, pursuing collaborative design of robot drive trains to explore real-life applications of ratios and proportions with societal and personal relevance.

2.b. Significance of using the TPACK framework

As the landscape of technological innovation continues to change rapidly, there has been a corresponding growth in the rate of cyclical obsolesce of prevailing educational technology and its replacement with ever newer technologies. This dynamics creates a demand on teachers to continually stay abreast with advances in technology [7]. However, most modern technologies are envisioned in response to business and industry needs and do not factor their promise for the education sector [7, 13]. Thus, teachers need guidance and support in examining the suitability of new technologies for promoting disciplinary learning among students and in building their own capacities for using technologies to address their pedagogical goals. In this spirit, an essential aspect of a technology-centered professional development program must be to create a supportive environment for teachers to interactively experience authentic instantiations of modern technologies in relevant disciplinary and pedagogical contexts. Successful incorporation of technology for the dual purposes of learning STEM disciplines deeply and preparing for entry into a workforce requiring tech-savvy further exemplifies the requirement for supporting teachers to effectively intertwine technology in their pedagogical practice in pursuit of their disciplinary goals. Unfortunately, due to their lack of formal exposure to, preparation in, and experience with advanced technologies, many teachers remain averse to applying these tools in their classrooms [14, 15].

Such unfamiliarity with technology and reluctance in its adoption demand for professional development programs that embed experiential explorations interweaving technology, pedagogy, and content to cultivate teachers' TPACK and self-efficacy (i.e., perceptions about one's ability to arrange and enact behaviors that are essential to attain specific performance levels [16]). Studies examining the effect of teachers' TPACK self-efficacy for robotics-based lessons have demonstrated a correspondence between TPACK self-efficacy and student performance. For example, the study of [17] focused on two teachers teaching three robotics-based science and math lessons and, according to the study results, higher TPACK self-efficacy scores of teachers were suggestive of favorable performance of their students. Thus, professional development programs that specifically seek to enhance the TPACK self-efficacy of teachers can assist them in becoming proficient practitioners of technology integration in the classroom and, thus, engender deeper learning of disciplinary matters among students.

Since the intentional development and exercise of TPACK imparts clarity and deeper understanding among its practitioners, this framework can be adopted as a methodology to design, assess, and refine various technology-based educational intervention [18], for example, teacher education programs and STEM curriculum for varied groups of learners, including K-12 students, pre-service teachers, in-service educators, etc. In addition, adopting the stance of TPACK for formulating and conducting education research, for example, in teacher professional development or design of technology-integrated curriculum, can help catalyze valuable recommendations and test hypotheses, which can be adopted by educators to acquire greater adeptness in technology use for supporting learning *versus* its superficial use as a vacuous tool [6]. In this spirit, this chapter illustrates applications of the TPACK framework in building the TPACK self-efficacy of teachers through professional development and designing STEM curriculum from a TPACK perspective.

2.c. Criteria for effective integration of technology as a learning tool

The process of selecting and assessing a technology for its effectiveness in classroom use can represent an intricate undertaking, especially for teachers who are not well-versed with its relevance and implications for teaching and learning [6]. Earle [10] and Ferdig [11] have both suggested that all technology tools considered for educational purposes must engender student-centered environments, foster active learning, and render pedagogically sound innovations. Collectively, the criteria advanced by Earle and Ferdig can serve as a helpful checklist in assisting teachers to design effective technology-integrated lessons. In particular, according to Earle [10], the integration of a technology in a lesson is judged pedagogically warranted, if the technology:

- enables novel ways of teaching and knowing that are unrealizable in its absence;
- provokes a sophisticated treatment of concepts;
- embeds greater opportunities for students to interact with the disciplinary content;
- sparks interest for teaching and learning among teachers and students, respectively; and
- relieves time for quality interactions in the classroom.

Analogously, Ferdig [11] advocates that a technology be acknowledged as pedagogically effective, if the technology:

- has attributes that foster it to be used appropriately, that is, the technology used is conducive to the process of teaching and learning;
- profits content learning outcomes, that is, the technology impacts student learning of the targeted subject matter beyond learning the use of technology itself; and
- supports varied methods for comprehensive and sophisticated analyses, that is, the technology embeds means to observe and acquire data concerning social and emotional outcomes. Moreover, the technology ought to facilitate the measurements of both affective (social and emotional) and cognitive gains.

It is not mandatory that a technology tool being examined for instructional use meet all of the criteria enumerated above. Instead, some or all of the aforementioned criteria must be weighed along with other relevant factors, for example, perceived familiarity with and ease of use for technology among students; sense of level of comfort for technology among teachers; availability of technology in required numbers for deployment; classroom context; etc. An educator must seek to fulfill as many of the above criteria as possible to create technology-supported experiences with meaning and purpose.

3. Literature review on teachers' TPACK development

Since its inception, the TPACK model of interactions among the underlying knowledge domains has been extensively deployed to prepare teachers in creating technology-infused lessons that effectively treat content cutting across disciplines and grade levels. Several research studies have examined the role of teacher preparation programs in building and measuring TPACK among teachers and proposed principles that can have significant implications for technology-integrated learning. For example, a careful analysis and synthesis of results from several TPACK studies has revealed that TK of teachers is related to their TPACK [15]. This of course suggests that teachers ought to have opportunities for exploring various features and operational modes of technology tools to attain confidence and, eventually, become motivated to integrate the selected technology for teaching and learning. According to [19], some promising strategies for improving teachers' TPACK include: modeling the integration of technology with pedagogy and content as well as cultivating opportunities for active engagement in creating and examining technology-integrated lessons. Next, we briefly review, as exemplars, two teacher preparation programs that focused on building TPACK for teachers.

The first example concerns a longitudinal study [20] for assessing the TPACK development of preservice teachers as the outcome of an 11-month-long graduate education program that entailed elaborate coursework delivered over a span of three semesters. Specifically, the program included foundational courses on educational theories, education research frameworks, instructional methods, curriculum design, learning assessment, classroom management, etc. A key aspect of the program included a course on educational technology for improving learning. This course sought to introduce the preservice teachers to varied ways in which technology can support subject matter and instructional goals. Moreover, it included a capstone project in which they designed and developed their own lesson embedded with technology. This second-semester course was synergistically connected to teaching methods courses on curriculum and instruction, first of which was favorably sequenced in the second semester. Finally, the program included several teaching practicums and an internship, which engaged the preservice teachers in supervised teaching. As part of the program, they created and presented their teaching e-portfolios.

For program assessment, one primary data source consisted of preservice teacher responses to the TPACK survey of [21] at four different times. Specifically, the TPACK survey was administered at the start of the first semester and subsequently at the end of each semester during the three-semesters-long graduate program. Other data sources included preservice teachers' reflection statements, their lesson plans, and their responses to an interview. The data analysis revealed that the largest growth in TPACK was realized during the second semester, most likely since the preservice teachers were then enrolled in the educational technology course. The course had allowed them the opportunity to apply their newly acquired knowledge to develop their own technology-infused lesson and obtain feedback to improve it. Subsequently, in the third semester, no growth in preservice teachers' TPACK was evidenced. Nonetheless, in this later period, their TPACK did not see any decline either, since they remained confident and motivated in applying technology in their teaching methods course and in teaching practicum. The data analysis further revealed that TPK and TCK domains of TPACK may not develop concurrently and may depend on the goals of the coursework. The study suggests that it is critical to examine TPACK development in teacher education programs over time and, thus, make necessary amends to render the program meaningful.

Another study [22] sought to examine the TPACK development among preservice English language teachers as an outcome of a five-week-long TPACK training workshop. The 59 participants in the workshop were in the third year of an English Language Teaching (ELT) academic program and were concurrently enrolled in a teaching methods course focused on ELT. The assessment data included pre-/post-workshop responses of workshop attendees to a TPACK survey and their journal entries documenting their experiences with educational technology for ELT as well as their views about integrating technology skills, instructional abilities, and English language knowledge for ELT effectively. The pre-workshop assessment data reveals that the attendees could enumerate myriad digital hardware and software for possible adoption in the classroom. Nonetheless, concerning technology infusion in ELT, they had ill-formed notions and could ideate only naïve uses, without any meaningful regard for learning goals. During the workshop, the attendees learned about several educational technology tools and strategies for their purposeful adoption to contribute to content learning goals for students. The post-workshop assessment data indicates that the attendees could name many digital applications, websites, and software tools with a high-degree of relevance for educational settings. They offered meaningful uses of technology tools for ELT. Moreover, they had a heightened level of awareness to effectively embed technology in ELT, for example, to enhance student motivation and learning gains. The TPACK score of attendees was seen to have increased with a concomitantly positive effect on the CK, PK, and PCK domains. The sample ELT materials developed by the workshop attendees illustrated their enhanced capacity to tailor English language lessons for serving intended learning goals by effectively interweaving their newly acquired understanding of technology tools with their knowledge of content and pedagogy. Drawing lessons from such pre-service education programs is vital for those who champion effective teaching and learning of disciplinary content with technology.

4. Development of teacher TPACK through professional development aimed at using robotics as a learning tool

To promote the purposeful infusion of technology in classrooms, it is essential that teachers receive effective professional development on technology use for improving student learning of disciplinary content. For using the construct of TPACK productively, teachers must possess the knowledge of the disciplinary content, a deep understanding of the intended technology, and a capacity to shape teaching and learning with practical uses of technology. In this section, we illustrate findings from research [9, 23] on: the evolution in the TPACK for a cohort of teachers through a professional development workshop; TPACK self-efficacy for teachers using robotics as a learning tool in classrooms; and recommendations to improve teachers' TPACK self-efficacy.

4.a. Details of a professional development workshop using the TPACK perspective [9]

The professional development workshop was planned as a three-week-long design-based experience [24] to be conducted under the mentorship of a collaborative team of engineering and education researchers. Four middle school science and math teacher participants of the workshop experientially engaged in a standards-aligned draft STEM curriculum that sought to alleviate content-specific instructional challenges through the integration of robotics. In advance of the workshop, the research team had collaboratively created the robotics-based STEM curriculum consisting of five science and five math lessons by utilizing the TPACK framework with attention to the role of the robot while catering to the known instructional needs for specific content elements. During the workshop, under the umbrella of TPACK, the teachers were prompted to reflect on the struggles of students with the concepts treated in the 10 lessons; examine, digest, analyze, and refine the teaching and learning strategies of these lessons; and project their corresponding outcomes. To achieve this, the teachers were introduced to the premise of each lesson with an introductory presentation and then they were encouraged to perform the related hands-on robotics activities. Next, after gaining familiarity with the TPACK framework and the technology integration criteria of Earle [10] and Ferdig [11], the teachers deliberated on the deficiencies of traditional ways of teaching the lesson's content and participated in a reflective dialogue. Following a proactive and engaging approach, the teachers documented whether the lesson satisfied its intended learning objectives while using robotics as a technology tool. Teacher reflections portraying myriad positive aspects of three robotics-based lessons on proportional relationship, center of mass and modeling, and statistics are detailed in [9]. Alternatively, the teachers reflected on whether the technology use in a lesson was divorced from pedagogy, for example, see corresponding details for a geometry lesson below. This program design was intentionally adopted as a strategy to develop teachers' TPACK through modeling, embed opportunities to revamp the curriculum [19], and provoke reflection on their learning.

After some exposure to the workshop's framing, lessons, and experiential learning activities, the teachers identified a geometry-based lesson, designed by the research team, constituting the case of a superfluous integration of robotics [9]. This lesson was created to expose to students that 3D objects are composed of layered 2D shapes. For example, a 3D cube results from layers of 2D squares assembled on top of one another. For hands-on learning, a mobile robot instrumented with a marker is placed on a cardboard surface. The robot is directed to move along the perimeter of a specified 2D figure (e.g., a rectangle), tracing the same shape on the cardboard. Having traced several rectangles, scaled variously relative to the first one, the students are asked to cut the 2D drawings and build a 3D object by arranging the cutouts on top of one another. For example, the students can place the largest rectangle at the bottom and place other rectangles in a descending order of size to construct a rectangular pyramid. This lesson was intended to provide a visual representation of the topic to students. While the lesson met the content learning standards, the teachers identified that it failed to fulfill Earle's second criteria (i.e., sophisticated treatment of concepts) and Ferdig's first criteria (i.e., appropriate use of technology). Specifically, they reasoned that students can themselves scale and draw various 2D shapes and having a robot do it for them denies students such learning experience. This narrative reveals the strength of the TPACK framework in shaping teachers' constructive feedback concerning a vacuous application of technology in teaching a disciplinary concept.

The above geometry lesson was eventually abandoned and not used in any classroom. Instead, with their freshly gained understanding of TPACK, strategies for effective technology integration, and sensitivity for learners' requirements, the teachers brainstormed to propose a new lesson for a different abstract topic that they had previously found to be pedagogically challenging. Specifically, they conceived, designed, prototyped, and tested an exciting and effective classroom activity using a mobile robot moving along a number line to visually represent the process of adding and subtracting with positive and negative integers (see Figures 6.2 (a)–(b)). The teachers reasoned that the visual nature of the intervention will permit students to gain a deeper understanding of the concept and internalize the usually confounding sign convention in dealing with negative integers. In examining the use of robotics with the technology-effectiveness criteria, the teachers were enthusiastically positive about it as evidenced below. With the robot's use to visualize a mathematical process, which otherwise poses a conceptual

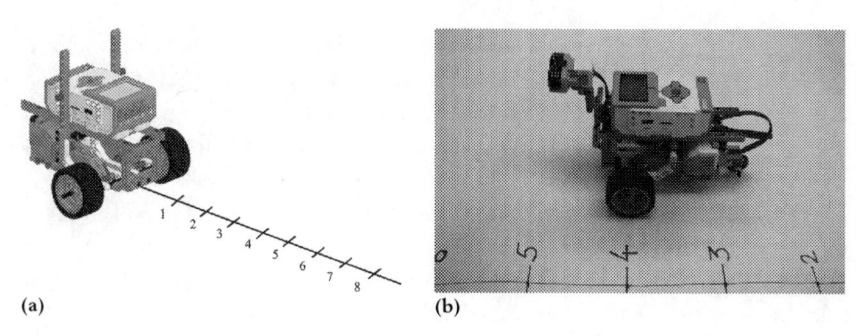

FIGURES 6.2 (a)–(b) Schematic (a) and experimental (b) representation of a robot with number line [9].

challenge to students, the teachers considered that the lesson represents an appropriate use of technology, thus passing Ferdig's first criterion. Even as the robot illustrates the process of adding and subtracting in an analogous way in which students are instructed on the topic, the teachers suggest that, by enabling students to numerically and visually validate the arithmetic process, sign convention, and result, this intervention satisfies Earle's first criterion: a novel way of teaching and knowing, realizable only with the robot. With students being prompted to enter the mathematical operation and two numbers on the LEGO brick and start the investigation by placing the robot at the origin of the number line, the teachers deem the lesson as fulfilling Earle's third criterion: embed greater opportunities for students to interact with the disciplinary content. Finally, in light of the positive evaluation of the lesson by the teachers, it is considered to meet Earle's fourth criterion: spark interest among teachers and students.

Throughout the professional development program, the teachers not only experienced and refined robotics-based STEM lessons, by adopting the stance of a critic, they assessed the role of technology in pedagogy. By drawing on their pre-existing knowledge of the disciplinary content of middle school science and math, practical experiences with teaching methods that would propel or impede student learning, and their newly acquired knowledge of robotics, the teachers swiftly abandoned a lesson plan wherein robotics would have symbolized a naïve technology integration and may have even hindered learning. Then, they collaborated to produce a novel lesson in which robotics was essential to engendering a new representation of intended content knowledge. Specifically, in this newly created lesson, the robot actively engages students in a learning process and obviates the need to memorize recipes, which the teachers characterized as problematic. With its ability to render a visually stimulating learning environment, this and other similar robotics-based lessons constitute a promising avenue of exploration for students with learning disabilities.

The above experience and additional reflections [9] of the teachers (concerning lessons on proportional relationship, modeling, and statistics) highlight their increased understanding of the TPACK framework and its appropriate application in a classroom setting. All of their reflections were studied in accordance with Earle's [10] and Ferdig's [11] criteria of effective technology integration. This example case study illustrates that a formal introduction to the TPACK framework and practical examination of illustrative technology-infused lessons using the criteria of Earle and Ferdig can enhance teacher capacity and skills in developing their own technology-integrated lessons. Supporting teachers in the effective implementation of resulting lessons can additionally increase student engagement with abstract STEM concepts and learning outcomes. Thus, these ideas should be essential components of a technology-focused professional development program.

4.b. Teachers' TPACK self-efficacy and recommendations to improve it using robotics as a technology tool

Understanding the influence of self-efficacy perception can serve an important role in developing capabilities, attaining better performance, staying motivated, and lowering anxiety [25]. Self-efficacy research concerning teachers and teaching has uncovered myriad ways in which personal efficacy beliefs can influence teacher effectiveness. Teachers who perceive higher self-efficacy exhibit a greater propensity to transfer new skills acquired from professional development to classroom; possess a superior capacity to explore novel methods for teaching and learning without any apprehension; test, refine, and sustain techniques that improve the likelihood of better learning outcomes; and contribute to an improvement in the overall school performance [25]. A recent research [23] investigated teachers' understanding of the TPACK construct, relative importance they assigned to various TPACK domains, and their TPACK self-efficacy regarding the use of robotics as a tool for teaching and learning. Twenty middle school science and math teachers who had previously attended a three-week-long robotics-infused STEM professional development workshop participated in the study. For this study, classroom enactment of lessons and experiential learning activities of students were observed as each teacher individually implemented a robotics-based science or math lesson in their classroom. Following the implementation of their lessons, the teachers responded to anonymous instruments that included a TPACK awareness survey and a TPACK self-efficacy survey (see sample survey items of [23] in Section 3 of Appendix A).

In the TPACK awareness survey, in one question, the teachers were asked to indicate the relative importance they accord to the TK, PK, and CK domains for effective planning and delivery of lessons using robotics. Based on the 17 responses, the teachers perceived TK, that is, the knowledge of different components, features, and operations of robots, as the most essential to teach the robotics-based science and math lessons. This outcome is indicative that, as recent adopters of robotics for the classroom, the teachers must have experienced myriad construction, assembly, programming, and troubleshooting burdens. Next, the respondent ranked CK domain as second in importance and lastly the PK domain. With the inclusion of robotics activities, as the teachers sought and delivered tangible representations of disciplinary content, they may have perceived pedagogy to have a lesser implication in this setting. Nonetheless, the pedagogical domain may additionally be influenced by the amount of time devoted for planning different aspects of a lesson, student requirements, etc. Thus, the lowest relative importance assigned to PK in this study does not diminish the important role of various factors of PK that contribute to the design and delivery of an effective robotics-based STEM lesson.

The teacher TPACK self-efficacy for robotics lessons was examined using a survey instrument that was adapted from a validated protocol [21]. Among the seven knowledge domains of TPACK, mean self-efficacy scores revealed that the 16 responding teachers perceived having the highest self-efficacy in their CK domain followed by PK, a result that can be ascribed to their sound knowledge of disciplinary content in a teaching context. The respondents had the least mean self-efficacy for TK domain due to their relatively new experience with robotics and less familiarity with troubleshooting-related technical difficulties. Finally, the survey responses revealed that the overall TPACK self-efficacy of teachers was not adequately high.

In response to the outcomes of TPACK self-efficacy survey [23], a subsequent case study [17] focused on providing two teachers special support to prepare for robotics-based science and math lessons and to determine skills and knowledge required for those lessons. The case study [17] also included a quiz to determine student learning outcomes after engaging with robotics-aided lessons. Following the analysis of results, a relationship was extrapolated between teacher self-efficacy and student learning outcomes. Thus, consistent reflection on their own TPACK and student improvement led the teachers to build their confidence levels in implementing robotics lessons and, in turn, enhancing student outcomes. The result of the analysis of [17], that is, the relationship between TPACK self-efficacy and student learning outcomes, encapsulates the promise of technology for teaching and learning. Moreover, the framing of a favorable connection between the TPACK self-efficacy and student learning outcomes can be appropriately adapted and applied by educators and school administrators to design and conduct professional development programming as well as construct and sustain professional learning communities.

Next, based on the findings of the study of [23], the following strategies have been recommended for teachers and schools to excel in building the TPACK competence for themselves and among their teaching staff, respectively. Teachers must occasionally partake in professional development opportunities specific to TPACK to both develop and sustain their self-efficacy [26]. Such a professional development program must prompt teachers to model the use of technology tools to support pedagogical goals and also engage them in hands-on experiential activities during a simulated classroom to experience the use of technology tools for enhancing learning [15, 27]. In addition to attending professional development programs, teachers should collaborate with their peers in the school and technology experts, participate in learning communities with other teachers to collaboratively solve challenges, and engage in self-study [23, 27]. The schools should create a supportive setting that encourages teachers to explore and integrate the TPACK framework in their own teaching. Among other activities, the schools can organize TPACK workshops, provide resources to sustain a TPACK learning community, and create mechanisms and structures for sharing best practices and addressing concerns.

In planning and executing lessons, teachers can experiment with varied levels of technology integration and create avenues for student engagement [23, 27]. For example, it may be productive to use technology for hooking learners to a topic, introducing new terminology related to the technology, conducting formal assessment, elaborating on a concept, etc. This can also include familiarizing with different features of the technology as well as knowing the affordances and challenges of a technology concerning classroom needs. Teachers can additionally reflect and analyze the abilities and needs of the diverse student groups in preparing lessons for effective technology integration [17]. To develop teachers' TPACK, schools must provide suitable classrooms and lab facilities that adequately satisfy or exceed the technology integration requirements. If feasible, conducting periodic assessments of the state of technology integration in lessons and classrooms of teachers can help in measuring their TPACK proficiency [20, 23] and creating needed support.

5. Development of TPACK-guided robotics-based STEM learning units

Incorporation of robotics in a classroom context can produce excitement for and foster involvement in STEM learning for a broad range of students [28]. Robotics-based lessons help in making learning engaging and rendering abstract concepts more accessible to students [9]. Instructional units grounded in robotics have the capability to address an array of topics ranging from fundamental STEM ideas to advanced content such as automation and control [9]. Lessons with robotics present opportunities to treat disciplinary content through multiple modes of representation, making learning more inclusive to a wide variety of educational settings. Robotics-based lessons have been shown to foster collaborative learning and problem-solving skills in active learning environments [9]. Under the TPACK framework, robotics-based lessons can be formulated to provide hands-on learning experiences, give learners opportunities to construct and own their learning, and most importantly, motivate learners to acquire skills that will assist them in accomplishing their goals [8].

The two learning units (or lessons) discussed below demonstrate the effective integration of STEM concepts with robotics as the technology component of the TPACK framework. Each lesson is followed by an examination of the various TPACK components needed for its development as well as Earle's [10] and Ferdig's [11] criteria concerning the effective use of technology. Both lessons have been implemented in classrooms of urban schools by teachers who had prior professional development experiences on teaching STEM using robots. The first lesson on biological adaptation is targeted toward elementary school students and the second lesson on least common multiple (LCM) is aimed toward middle school students. The implementation of lessons was observed by engineering and education researchers in the classrooms to note their impact on student learning and to evaluate the lessons regarding the technology integration effectiveness criteria of Earle and Ferdig.

5.a. A learning unit on biological adaptation for elementary school students [4, 29]

Standard pedagogical approaches for teaching biological adaptation and illustrating its attributes to students appear to fall short. Based on the analysis of the understanding of biological adaptation among secondary school students, prior research has established that "students find the subject area difficult" and proposed the examination of alternative perspectives of the topic, in varied settings, to advance student understanding [30]. Introduction of this subject matter in the science curriculum of early elementary grades is another recommendation of [30]. As one illustration of biological adaptation, locomotion mechanisms that are specific to a certain species can be explained in a classroom. Unfortunately, using traditional pedagogy, it may be challenging for students to visualize and experience the role of various characteristics of biological adaptation that give rise to the prevalence or elimination of varied locomotion mechanisms and behaviors. Nonetheless, robotics offers a means to mimic many biological concepts in classroom teaching and learning. For example, in studying the species-specific locomotion mechanisms, students can alter various anatomical features of its robotics model to simulate their possible effect on the survival of the species with favorable biological attributes. In this spirit, a learning unit has been conceived, prototyped, and tested for enhancing student comprehension concerning the phenomenon and concept of biological adaptation. In particular, using a robot as the physical model of an arctic animal, students formulate and test hypotheses about the benefit (or penalty) of various locomotive features of the animal. The formulation of this unit under the inquiry-based framework is critical in allowing students to build their learning confidence and co-create their experiences with peers. See Table 6.1 for details.

TABLE 6.1 A biological adaptation learning unit for elementary school grades

Objectives: By the end of this lesson, students will be able to analyze different abilities and structures of certain animals and describe the biological adaptations in animal species in different environments.

Key questions: How does adaptation emerge among different animals in response to varied environments? What will happen to an animal species that fails to adapt with a change in its settings?

Performance assessment: Students will create a report consisting of the problem statement, a potential solution and its rationale, observations and test results, and conclusions.

Material required: For each group of three to four students: a LEGO robot, a roll of flexible metal mesh, a sticking tape, and a ruler or measuring tape. For the entire class: one big aluminum foil tray filled with cotton balls (stand-in for snow), two big aluminum foil trays each filled with difficult-to-navigate material, such as sand, marbles, toothpicks, etc.

Activity preparation: Write and upload a program using LEGO, or other available programming environment, that moves the motors of the robot forward for 30 rotations and then stops. Create a set of "shoes" for the robot using the flexible metal mesh. Fill one foil tray with cotton balls and organize them to create a one-cotton-ball-thick layer.

Learning unit:

- Begin with a discussion on the various features of different animal species that the students may have noticed in nature. Ask: what are some commonalities, what are some differences, and what are some peculiar features that make one species different from any other?
- Using a presentation, introduce the concept of biological adaptation to the students. Delve deep into any one preferred example. Share briefly about the robotics activity and the setup.
- ***Activity #1—Biological adaptation of a snow leopard*** with the walker robot of Figure 6.3(a):
  - o Place the walker robot on a flat tabletop and execute the program that drives its motors to move the robot forward on the table. Demonstrate that, with each press of the "run" button, the robot moves forward the same number of steps and then stops.
  - o Show to the students the cotton ball tray. Now, demonstrate the walker robot moving forward in the tray of cotton balls. Ask the students to observe any difference in performance compared to the earlier walk of the robot on the flat surface. Have the students report on the worksheet any problems encountered by the robot in the new environment.
- *Activity #2—Design Challenge*:
  - o Explain to the students the assignment to design and test a "shoe" add-on for the walker robot so that it can move effectively across the cotton balls (or sand, marble, etc.). See examples in Figures 6.3(b) and 6.4(a).
  - o Organize the class into several teams, each with three to four students. Direct them to use the internet for researching inherited adaptations for animals who thrive in snow environments (e.g., a snow leopard).

(*Continued*)

TABLE 6.1 (Continued)

- o Ask the students to brainstorm and suggest an idea for a "shoes" upgrade for the walker robot.
- o Explain that the mesh material costs $5 to make a 1 square cm shoe. Ask the students to discuss cost-benefit tradeoff, for example, increased shoe area may improve distance covered in 30 steps for a higher shoe cost. Clarify the design goal about the need to create shoes that produce the greatest walking distance per dollar expense.
- o Have team members sketch their shoe designs, select its shape, and then build the shoes.
- o Ask each group to compute the area (A) of the shoes. For rectangular shoes, use $A = L \times W$, where L and W denote length and width respectively. Alternatively, for circular shoes, use $A = (3.14 \times d^2)/4$, where d denotes the diameter. Finally, if the shoes are odd shapes, have the students trace the shoe shape onto a graph paper and count the number of squares occupied by the shoe perimeter. Record the shoe areas on the worksheet.
- o Next, have the student teams test their "adaptations" for the robot shoes by having the robot walker move in the tray of cotton balls (see Figure 6.4(b)) and make required modifications to their shoes design.

- ***Student showcase:*** Have teams take turns to demonstrate the experiment to their peers.
- ***Design report and reflections:*** Assign each student team to write a short test report consisting of: the problem statement, a potential solution and its rationale, observations and test results, and conclusions. Ask them to add their reflections.

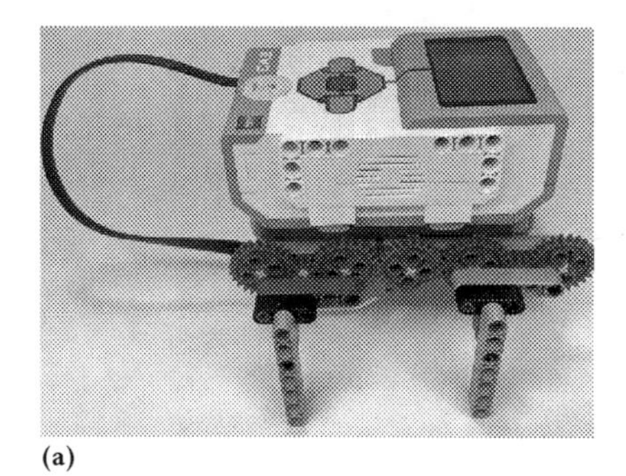

(a)

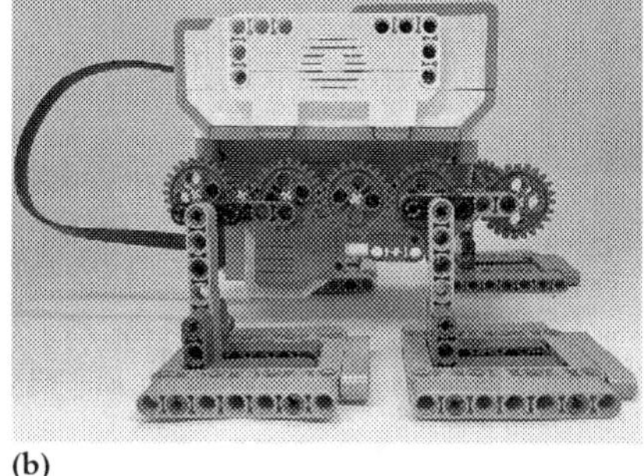

(b)

FIGURES 6.3 (a)–(b) A robot walker with pointed feet (a) *versus* wide-based feet (b).

(i) TPACK necessary to develop the learning unit on biological adaptation

This learning unit illustrates a robotics-infused, activity-based instruction for a biology topic at the elementary school level. As previously indicated, it is pedagogically difficult to create alternative perspectives and tangible representations of biological adaptation for classroom instruction and learning, thus justifying the consideration of the TPACK framework for treating this subject matter.

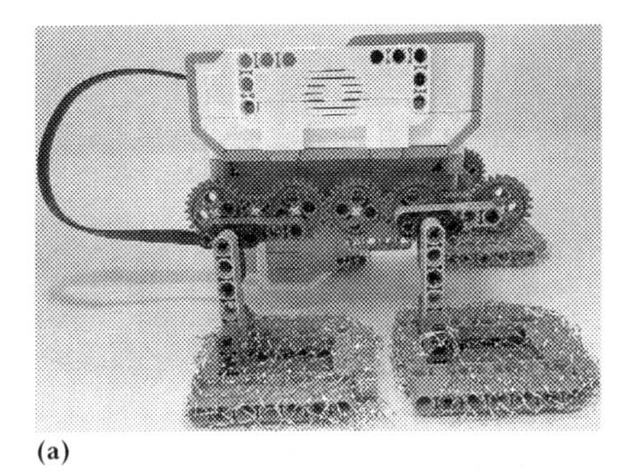

(a)

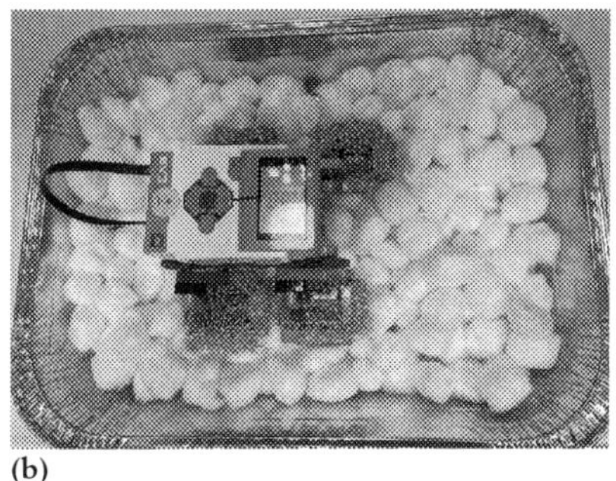

(b)

FIGURES 6.4 (a)–(b) A robot with wide-based feet covered using a metal mesh (a) and an experimental setup with the robot traversing in a tray of cotton balls emulating the snow environment (b) [29].

First, the development of this learning unit entails CK that includes the understanding of different abilities and structures of animals, inherited biological adaptation in animals, importance and attributes of biological adaptation, and ancillary concepts from other areas. For example, it is important to know the math concept concerning area for various shapes to experiment with varied shoe geometries that may be favorable biological adaptations for snow environments. Second, this unit requires PK of effective lesson planning, assessment planning, classroom management, inquiry-based learning, collaborative learning, and interdisciplinary teaching. Third, the inclusion of robotics and PowerPoint presentations in the unit necessitates the TK about these tools and their features, such as how to use various components of the robot, how to program it, and how to design engaging presentations using PowerPoint. Fourth, PCK to develop this learning unit entails an understanding of tasks, activities, and questions that can reveal and address student misconceptions. In fact, it requires knowledge and understanding to design activities that embed applications of content from another discipline in a scaffolded manner. For example, pedagogical strategies of collaborative group work and design challenges used in this unit promote creativity, analysis, and synthesis of disciplinary content from both biology and math. In the absence of a content-aligned technology intervention, learners may become passive consumers of knowledge and they may miss the opportunity to build their own understanding through experiential explorations. Fifth, in response to the above, TCK of this unit entails use of robotics that renders an alternative and material representation for the concept of biological adaptation and promotes its deep understanding through hands-on inquiry. Specifically, in this learning unit, the robotics structure and programming (i.e., TK) are used to explain biological adaptation for snow leopards (i.e., CK). Sixth, TPK in this unit employs myriad pedagogical strategies, for example, inquiry-based learning, collaborative group work, and interdisciplinary learning, for a purposeful integration of robotics for the main activity. Use of the walker robot, its program, and its walking environment all promote kinesthetic learning through hands-on experimentation and self-exploration. Moreover, PowerPoint presentation is used to engage all students with multiple modes of representation—such as text, images, and video.

Overall, in this unit, the TPACK framework supports the incorporation of robotics as a tool to promote inquiry-based learning for the topic of biological adaptation. The walker robot, serving as a physical analog of a snow leopard, promotes student engagement in inquiry learning by offering them a tangible representation, as a physical artifact, of biological adaptation. As students experientially explore the influence of their designed shoes for the robot feet, they begin to link the relationship between the ability of an animal to traverse its environment, its ability to find food for survival, and passing on favorable attributes as inherited features. The robotics activity allows students to visualize some factors that may support biological adaptation. Specifically, a robot with pointed feet cannot travel easily through the cotton balls compared to the robot with wide-base feet. As the differential ability for traversing in snow influences the ability to find food for survival, differentially, adaptation emerges from inherited traits that support survival. In summary, under the TPACK construct, this biological adaptation unit purposefully integrates the knowledge domains of content, pedagogy, and technology.

(ii) Effective integration of robotics as a learning tool

In contrast to activities that employ print- or presentation-based technologies, which may prompt passive consumption of knowledge, this activity gives students an opportunity to construct their own learning experiences. The lesson utilizes three technology tools that promote meaningful participation in learning. Robotics, the primary technology tool, meets all of the technology value-add and effectiveness criteria of Earle and Ferdig [10, 11]. Specifically, concerning Earle's criteria, the use of walker robots fosters novel teaching and learning experiences for the abstract topic of biological adaptation, enables sophisticated treatment of biological adaptation using hands-on explorations, integrates tasks and activities that encourage students to interact with and examine the subject matter, promotes interest through various inquiry tasks for teaching and learning, and embeds time for interactions among students through collaborative activities. Similarly, for Ferdig's criteria of effective use of technology, the robotics-based unit on biological adaptation is deemed to serve pedagogical needs appropriately by engaging students in the topic, conducting investigations at their own pace, and processing the ideas of adaptation deeply. Moreover, the unit contributes to student learning outcomes by affording them opportunities to examine, analyze, synthesize, and present their findings about animal adaptation. In fact, according to the pre- and post-assessment responses by students, the robotics-based lesson enhanced their understanding of the relationship between the inheritance of favorable adapted anatomical characteristics and survival of the animal.

Finally, this learning unit promotes social and emotional learning goals through the incorporation of collaborative learning and allowing students to invest interest in the survival of the animal (i.e., the walker robot). The two secondary technology tools, namely, the internet and PowerPoint software are used for conducting research and delivering multimedia presentations, respectively. These technology uses are purposeful and deemed to meet various effectiveness criteria. Overall, the three technology tools are meaningfully integrated in the learning unit which adds value in making learning relevant.

5.b. A learning unit on the application of least common multiple in a real-life scenario [31]

Similar to various other mathematical concepts and procedures, middle school teachers commonly teach the concept of least common multiple (LCM) verbally, in abstract, divorced from its real-world implications. After illustrating the recipe for computing the LCM, they assign various drill and practice exercises in which students repeat the computational recipe of LCM to internalize the same. Unfortunately, with such a teaching and learning strategy, students do not gain an understanding of the need for or utility of LCM for any practical problem-solving situations [17]. As one remedy to this challenge, this learning unit proposes a robotics-based experiential activity whose objective is to illustrate an application of the LCM of two whole numbers in a real-life scenario. Specifically, the activity challenges students to apply the LCM concept for examining a practical situation involving two metro trains (each represented using a LEGO robot, see Figure 6.5). Running on the same route, one metro train is designated local with frequent stops while the other metro train is designated express with less frequent stops. Two LEGO robots are preprogrammed to model the one local and one express train. Next, for a practical problem involving the local and express trains, by operating the given LEGO robots and observing their stopping patterns, students analyze the problem of determining the LCM of two whole numbers (see Table 6.2 for details). In this manner, using the robot activities, students learn and practice the concepts of LCM in a real-world context that they find understandable and meaningful. The success of this lesson relies on the assumption that the students are familiar with their city's metro train mass transit system and that they have learned the whole number system in prior grades.

(i) TPACK necessary to develop the learning unit on least common multiple

This lesson incorporated robotics-enriched activity-based pedagogy in the teaching of a math concept at the middle school grade level. The concept of LCM is abstract, that is, it lacks a concrete and accessible representation for

FIGURE 6.5 Using the movement and stopping of two LEGO robots (local and express trains) to understand the concept of least common multiple [31], the colored paper pieces (seen here in light/dark gray) help identify the locations where the robots stop temporarily.

students. Teachers find it challenging to conceive and illustrate varied representations for the LCM concept to promote student understanding of this topic. In response, robotics is envisioned as an instructional support for this learning unit and it requires mindful consideration of the various domains of the TPACK construct in its creation. First, the development of this unit requires CK of germane mathematical concepts, including whole numbers, number line, LCM,

TABLE 6.2 An application of the least common multiple (LCM)

Objectives: By the end of this lesson, students will be able to determine the least common multiple of two whole numbers by analyzing a scenario using LEGO robots.
Key question: What are some examples of real-life scenarios where the concept of least common multiple can be applied?
Performance assessment: In groups, the students will conduct experiments using the robots as models for the local and express metro trains. They will observe, record, and analyze the experimental data about the halting patterns of the two metro trains. They will present their findings in the form of a report. Then, they will be given exit tickets to check for their comprehension of the concept.
Material required: Measuring tape, markers, graph paper, and LEGO robots.
Activity preparation: Prepare and upload the program corresponding to this activity on the LEGO robots. Arrange all the material for each group of students.
Learning unit:

- ***Do now:*** Entry slip with five drill and practice questions on finding the LCM of two whole numbers.
- ***Introduce the scenario:*** You and your friend are going to go see a movie. Upon entering the busy metrorail station, you and your friend get separated. When a local metro train and an express metro train going toward your destination arrive simultaneously, your friend boards the local metro train while you board the express metro train. You now wish to figure out the first metrorail station at which both of you can meet so that you can travel from there to the theater together. What metrorail station will you both meet at?
- ***Activity #1—Data collection and analysis using robots:***
  - o Prepare two robots–one representing the local metro train and the other representing express metro train–to travel along straight lines parallel to one another.
  - o The local robot will travel forward and halt every 2 seconds while the express robot will travel forward and halt every 5 seconds. With the two robots starting simultaneously from the same origin, moving at the same speed, will they ever halt at the same station? If yes, then determine the first stations that the two robots will arrive at together?
  - o Make the students aware of the time display on the two robots.
  - o As the robots travel their path and halt at respective stations, mark each halt.
  - o Give the students a data table to fill out. The data table includes fields for the time elapsed for each station on the two robots.
  - o Ask the students to illustrate their findings regarding the timing of the two robot using two number lines.
  - o Ask the students to substantiate their results by using an algorithm to determine the least common multiple.
  - o Ask the students to showcase their findings, explain their algorithms/rationale, and present their understanding of the concept of LCM by coming up with additional real-life scenarios where the concept of LCM can be used.
- ***Assessment:*** Use a graph paper to address this scenario: "Commuter trains depart from a train station every 10 minutes. Metro trains depart every 4 minutes. A metro train and a commuter train both leave the station together. How long will it be before another metro train and commuter train depart from the station at the same time?"

data collection, and data analysis. Second, in addition to the deep knowledge of content, the unit also requires PK concerning lesson planning, assessment planning, classroom management, activity-based pedagogy, and collaborative learning. Third, to promote student engagement in technology-enriched math content learning, the unit requires TK of LEGO robotics and programming. Fourth, acknowledging the intangible nature of the topic of LCM, PCK that can best scaffold student learning of content includes the design of activities that enable learners to consider the implications of their new understanding for real-life scenarios and that promote collaborative problem-solving for personally meaningful situations. As one illustration of such PCK, designing a content learning activity that builds on the knowledge of students' life experiences (e.g., students in urban environments commute by metro trains and are familiar with local and express trains) offers a compelling approach to integrate real-life scenarios as practical applications of classroom learning. Fifth, to teach the concept of LCM with robots requires TCK concerning the appropriate manner for using LEGO robots to represent the intersection of a real-life scenario of metro trains with the math content. Having the knowledge of specific features of the robot and their suitable use to teach the math content is particularly relevant in making the concept of LCM tangible and accessible for visualization. This illustrates an effective leveraging of the affordances of robotics as a formidable learning tool. Sixth, the learning unit for LCM entails TPK wherein LEGO robots are employed in design and explorations activities that engender enthusiasm and promote engagement with peers. Moreover, TPK includes the use of robotics as a tool that embeds a real-life problem-solving scenario to make the concept meaningful to students and to deepen their knowledge. Finally, TPACK for this unit systematically integrates the three foundational constituents, that is, robots as a technological tool, activity-based pedagogy, and the math content of LCM. With such an integrated view of TPACK, teachers can encourage and engage students to collaboratively solve a problem concerning LCM by collecting and analyzing data using robots and in the process gaining an alternative and concrete representation of the LCM concept with robots. Following such an experiential exercise in practical problem-solving, teachers can support their students to extrapolate their learning to gain an improved comprehension of the LCM concept. This showcases that all the constituent elements of TPACK have been meaningfully interweaved to engender a deeper and novel learning experience.

(ii) Effective integration of robotics as a learning tool

The robotics-based hands-on activity for the LCM unit can trigger active learning instead of making students perform drill and practice calculations that fail to generate any understanding of, are devoid of any valid rationale for, and divorced from any real-life implications for the concept of LCM. Employing Earle's [10] criteria for the value brought by technology in a classroom, the use of robots to model local and express metro trains in an activity-based context, which encapsulates the concept of LCM for the practical problem of finding a common stopping station for the two trains, constitutes a novel mechanism for teaching and learning this math topic. Use of local and express robots provokes a deep processing of concepts since it enables an experimental visualization of the LCM of two whole numbers. Such an experiential process of learning with robotics transforms the abstract idea of LCM into an opportunity to interact with, learn, and understand this idea in a tangible manner. The lesson supports collaborative peer work that promotes student interaction with each other; generates enthusiasm for teaching and learning; and engenders opportunities for classroom demonstrations, presentations, and discussions. Evaluating against Ferdig's [11] criteria of effectiveness of technology in pedagogy, it is evident that the LCM unit utilized robotics appropriately to foster active learning experiences that may have been challenging to create with alternative technologies. Furthermore, the incorporation of robotics-based activity, data collection, observation of stopping patterns of two robots, and data analysis all provided varied ways for deeper learning, understanding, and assessment of content learning outcomes. Finally, the use of multiple robots and assigning varied roles to students facilitated collaborative teamwork that offered possible avenues for their social and emotional development. To summarize, the use of robots in the LCM lesson provides an opportunity to enhance the quality of student interaction with the topic by infusing excitement and enthusiasm along with diverse means to comprehend the concept. All the elements seamlessly connect together to produce novel, relevant, and deeper ways to construct learning, and thus meet all of the value-added and effective technology integration criteria of Earle and Ferdig.

6. Conclusion

Use of the TPACK framework can help educators to focus on the appropriate use of technology tools, in alignment with the targeted area of curriculum, to facilitate effective instruction. Integration of technology for learning should go beyond creating a novelty effect, which despite engendering excitement in the classroom may fail as a pedagogical device for producing accessible exemplars of difficult concepts. Specifically, it must foster deeper processing of disciplinary concepts, attend to diverse learning needs of students, and enhance the quality of classroom interactions. Moreover, the selected technology tool must meaningfully connect and cater to the overall learning goal. The realization of this vision of technology integration for learning can be fulfilled through TPACK-centered professional development programs that support teachers in developing self-efficacy, confidence, comfort, knowledge, and skills for implementing effective technology interventions in their classrooms. Furthermore, the development of technology-based learning units, such as the two lessons showcased in this chapter, constitutes another compelling strategy for meaningful technology integration. Adoption and application of Earle's and Ferdig's criteria in examining technology-based learning units can benefit teachers in determining the appropriateness and effectiveness of technology integration. In addition, the TPACK framework can help ensure that the selected technology is conducive to a structured and efficient integration with content and pedagogy. Overall, exploiting the inherent affordances of technologies, in service of effective teaching and learning of disciplinary content, can lead to active, relevant, engaging, and inclusive learning experiences.

7. Key takeaways

- TPACK is a framework that characterizes the varied knowledge domains essential for effective integration of technology with disciplinary content and classroom pedagogy.
- TPACK is composed of the following seven components whose acquisition and purposeful application can help teachers in crafting their lesson: content knowledge, pedagogical knowledge, technological knowledge, pedagogical content knowledge, technological content knowledge, technological pedagogical knowledge, and technological pedagogical content knowledge.
- Earle suggested five criteria to judge the value of technology in a lesson. The technology: enables novel experiences in teaching and learning that are infeasible without it, triggers a sophisticated treatment of concepts, embeds greater opportunities for students to interact with the disciplinary content, sparks interest for teaching and learning among teachers and students, and relieves time for quality interactions in the classroom [10].
- Ferdig suggested three criteria for evaluating the effectiveness of technologies in pedagogy: suitable use of technologies, influence on content learning, and diverse methods for deep and complete analyses of affective and cognitive gains [11].
- Robotics-based lessons embed opportunities to experience the targeted content tangibly through varied modes of representation, making learning more inclusive to a wide variety of educational settings. Such lessons create active learning environments wherein experiential activities with robotics catalyze collaborative learning and problem-solving in real-life practical contexts.
- Comprehensive analysis of several prior TPACK research studies has revealed a relationship between teachers' TK and TPACK [15, 19].
- Modeling the integration of technology with content and pedagogy and cultivating opportunities to partake in analyzing and refining the technology-integrated lessons serve as two promising strategies for improving teachers' TPACK [19].
- There is a direct relationship between teachers' TPACK self-efficacy and student performance, with higher self-efficacy being indicative of better student performance [23].

References

1. Ottenbreit-Leftwich, A.T., Glazewski, K.D., Newby, T.J., and Ertmer, P.A., Teacher value beliefs associated with using technology: Addressing professional and student needs. *Computers & Education*, 2010.55(3): p. 1321–1335.
2. Papert, S., A critique of technocentrism in thinking about the school of the future, in *Children in the Information Age*, B. Sendov and I. Stanchev, Editors. 1988, Oxford, UK: Pergamon. p. 3–18.
3. Harris, J., Mishra, P., and Koehler, M., Teachers' technological pedagogical content knowledge and learning activity types: Curriculum-based technology integration reframed. *Journal of Research on Technology in Education*, 2009.41(4): p. 393–416.
4. Brill, A.S., Listman, J.B., and Kapila, V., Using robotics as the technological foundation for the TPACK framework in K-12 classrooms, in *Proceedings of ASEE Annual Conference and Exposition*. 2015, Seattle, WA; Available from: https://peer.asee.org/25015.
5. Shulman, L.S., Those who understand: Knowledge growth in teaching. *Educational Researcher*, February 1986.12(2): p. 4–14.
6. Koehler, M.J. and Mishra, P., What is technological pedagogical content knowledge? *Contemporary Issues in Technology and Teacher Education*, 2009.9(1): p. 60–70.
7. Mishra, P. and Koehler, M.J., Technological pedagogical content knowledge: A framework for teacher knowledge. *Teachers College Record*, June, 2006.108(6): p. 1017–1054.
8. Eguchi, A., Bringing robotics in classrooms, in *Robotics in STEM Education*, M.S. Khine, Editor. 2017, Cham, Switzerland: Springer. p. 3–31.
9. Brill, A.S., Elliot, C.H., Listman, J.B., Milne, C.E., and Kapila, V., Middle school teachers' evolution of TPACK understanding through professional development, in *Proceedings of ASEE Annual Conference and Exposition*. 2016, New Orleans, LA; Available from: https://peer.asee.org/25720.
10. Earle, R., The integration of instructional technology into public education: Promises and challenges. *Education Technology Magazine*, 2002.42(1): p. 5–13.
11. Ferdig, R.E., Assessing technologies for teaching and learning: Understanding the importance of technological pedagogical content knowledge. *British Journal of Educational Technology*, 2006.37(5): p. 749–760.
12. Wadsworth, B.J., *Piaget's Theory of Cognitive and Affective Development: Foundations of Constructivism*. 5th ed. 1996, White Plains, NY: Longman Publishing.
13. Zhao, Y., What teachers need to know about technology? Framing the question, in *What Should Teachers Know About Technology? Perspectives and Practices*, Y. Zhao, Editor. 2004, Greenwich, CT: Information Age Publishing. p. 1–14.
14. Moorhead, M., Elliott, C.H., Listman, J.B., Milne, C.E., and Kapila, V., Professional development through situated learning techniques adapted with design-based research, in *Proceedings of ASEE Annual Conference and Exposition*. 2016, New Orleans, LA; Available from: https://peer.asee.org/25967.
15. Wang, W., Schmidt-Crawford, D., and Jin, Y., Preservice teachers' TPACK development: A review of literature. *Journal of Digital Learning in Teacher Education*, 2018.34(4): p. 234–258.
16. Bandura, A., *Self-efficacy: The Exercise of Control*. 1997, New York, NY: W. H. Freeman and Company.
17. Mallik, A., Rahman, S.M.M., Rajguru, S.B., and Kapila, V., Fundamental: Examining the variations in the TPACK framework for teaching robotics-aided STEM lessons of varying difficulty, in *Proceedings of ASEE Annual Conference and Exposition*. 2018, Salt Lake City, UT; Available from: https://peer.asee.org/30550.

18. Baran, E., Chuang, H.-H., and Thompson, A., TPACK: An emerging research and development tool for teacher educators. *Turkish Online Journal of Educational Technology-TOJET*, 2011.10(4): p. 370–377.
19. Voogt, J., Fisser, P., Pareja Roblin, N., Tondeur, J., and van Braak, J., Technological pedagogical content knowledge–A review of the literature. *Journal of Computer Assisted Learning*, 2013.29(2): p. 109–121.
20. Hofer, M. and Grandgenett, N., TPACK development in teacher education. *Journal of Research on Technology in Education*, 2012.45(1): p. 83–106.
21. Schmidt, D.A., Baran, E., Thompson, A.D., Mishra, P., Koehler, M.J., and Shin, T.S., Technological pedagogical content knowledge (TPACK) the development and validation of an assessment instrument for preservice teachers. *Journal of research on Technology in Education*, 2009.42(2): p. 123–149.
22. Ersanli, C.Y., Improving technological pedagogical content knowledge (TPACK) of pre-service English language teachers. *International Education Studies*, 2016.9(5): p. 18–27.
23. Rahman, S.M.M., Krishnan, V.J., and Kapila, V., Exploring the dynamic nature of TPACK framework in teaching STEM using robotics in middle school classrooms, in *Proceedings of ASEE Annual Conference and Exposition*. 2017, Columbus, OH; Available from: https://peer.asee.org/28336.
24. The Design-Based Research Collective, Design-based research: An emerging paradigm for educational inquiry. *Educational Researcher*, 2003.32(1): p. 5–8.
25. Bray-Clark, N. and Bates, R., Self-efficacy beliefs and teacher effectiveness: Implications for professional development. *Professional Educator*, 2003.26(1): p. 13–22.
26. You, H.S. and Kapila, V., Effectiveness of professional development: Integration of educational robotics into science and math curricula, in *Proceedings of ASEE Annual Conference and Exposition*. 2017, Columbus, OH; Available from: https://peer.asee.org/28207.
27. Adipat, S., Developing technological pedagogical content knowledge (TPACK) through technology-enhanced content and language-integrated learning (T-CLIL) instruction. *Education and Information Technologies*, 2021.26(5): p. 6461–6477.
28. Mosley, P. and Kline, R., Engaging students: A framework using LEGO® robotics to teach problem solving. *Information Technology, Learning, and Performance Journal*, 2006.24(1): p. 39–45.
29. Cave, A., *Arctic Animal Robot*; Available from: www.teachengineering.org/activities/view/nyu_arctic_activity1.
30. Engel Clough, E. and Wood-Robinson, C., How secondary students interpret instances of biological adaptation. *Journal of Biological Education*, 1985.19(2): p. 125–130.
31. Rahman, S.M.M., Chacko, S.M., and Kapila, V., Building trust in robots in robotics-focused STEM education under TPACK framework in middle schools, in *Proceedings of ASEE Annual Conference and Exposition*. 2017, Columbus, OH; Available from: https://peer.asee.org/27990.

PART III

Instructional perspectives and lesson designs

The final part of the book showcases the alignment between the theoretical perspectives and their translation toward implementation in classroom settings. Specifically, through three chapters, this part introduces the readers with perspectives, frameworks, plans, and implementation ideas that can help them in planning for their own classrooms. The seventh chapter of the book, *Prerequisites, Practices, and Perceptions to Design Effective Robotics-based Lessons*, introduces the readers to the necessary viewpoints and practices that can facilitate the implementation of robotics in classrooms and make the integration more effective. Next, the eighth chapter of the book, *Applying Cognitive Levels of Bloom's Taxonomy to Robotics-based Learning*, introduces the theory and examples of designing lessons that can render higher-order cognitive engagement among students. Finally, the ninth and last chapter of the book, *Using the 5E Model to Develop Robotics-based Science Units*, provides theoretical underpinnings and applications of the 5E instructional model in engendering conceptual change among learners through robotics-enhanced science units.

DOI: 10.4324/b23177-9

FIGURE PIII.1 Classroom implementation of educational robotics.

7

PREREQUISITES, PRACTICES, AND PERCEPTIONS TO DESIGN EFFECTIVE ROBOTICS-BASED LESSONS

1. Introduction

The myriad benefits of educational robotics in the teaching and learning of science, technology, engineering, and mathematics (STEM) are attracting increased interest from K-12 educators [1]. Incorporating robotics in K-12 STEM education contributes to developing cognitive and social skills for students [1]; making abstract concepts accessible through easy visualizations [2]; providing hands-on learning opportunities [3]; cultivating student engagement, motivation, and excitement [2, 4]; and improving the overall learning environment [4]. Nonetheless, the mere use of robots in a lesson doesn't invariably guarantee that all learning activities will lead to positive outcomes; instead, it is paramount to thoughtfully incorporate instructional practices that engender active engagement for students in robotics-based learning. An authentic and well-planned implementation is a key contributor to the success of robotics-based lessons.

Productive use of robotics in the classroom necessitates consideration of students' current knowledge *vis-à-vis* the prerequisites required for successfully learning through robots. Specifically, surfacing prerequisite knowledge and supporting students based on their prevailing understanding can aid in mitigating their anxiety, identifying their interests, and formulating instruction that meets their needs [5]. Similarly, to foster deep learning, effective instructional practices must allow the use of robotics as an experimental device to explore ideas, clarify misconceptions, and illustrate abstract concepts through real-world problem solving [6]. In addition to the instructional practices and prerequisite knowledge, student perceptions toward robots as learning tools represent another fundamental factor in the success of robotics integration [7]. Thus, the knowledge of teachers about student perceptions and factors that contribute to those perceptions is essential

DOI: 10.4324/b23177-10

to formulate robotics-based lessons that promote engagement, excitement, and maximize benefits [8]. Collectively, prerequisites for using robots for learning, instructional practices leading to active learning, and understanding of student perceptions toward robots constitute the three fundamental factors that contribute to making robotics-enabled learning in classrooms successful.

This chapter begins by highlighting the attributes that contribute to effective integration of robotics in classrooms alongside supporting studies from relevant literature. Next, the chapter presents three studies that were conducted in middle school classrooms in the northeastern United States (US) and that investigated the aforementioned three factors integral to producing effective learning in classrooms infused with robotics-based STEM lessons. The outcomes of the studies are accompanied with recommendations that can be employed in classrooms leading to the success of robotics-based learning.

2. Enabling effective integration of robotics in classrooms

Teaching and learning through robotics can be an unconventional process that does not proceed along a straight path. Robots as manipulatives engage students in constructionist and collaborative processes to make learning engaging and personally relevant [2]. Bers [9] recommends four attributes of a constructionist learning environment that constitute the foundation of effective educational robotics experiences. First, the learning environment should foster opportunities for students to explore and pursue their own ideas and projects and share the new knowledge, skills, and products with their community. Second, the robotics artifacts should be used to promote conceptual understanding by illustrating accessible and tangible representations of abstract ideas. Third, the robotics-enabled curriculum should cultivate a deep understanding of concepts, with robotics serving as a tool that delivers the concepts in an instructionally aligned manner. Finally, the robotics-infused learning environment should promote self-reflection through documentation and sharing.

To operationalize these foundational attributes of constructionist learning, it is essential to: recognize students' prior knowledge about concepts that are necessary for successful robotics-based learning, utilize instructional practices that generate active learning environments, and shape student perceptions that positively contribute to their motivation for studying STEM using robotics. Since the prevailing K-12 curriculum is largely devoid of opportunities that can prepare students to effectively participate in robotics-based learning activities, their success may be impeded in such endeavors at first [5]. Thus, by gaining a familiarity with the state of students' current knowledge, teachers can adequately tailor the curriculum and timely introduce new concepts and skills that must precede and support a robotics-based lesson. Prior research has highlighted the role of computational thinking [10] as a prerequisite to myriad aspects of STEM learning, such as programming [11], problem solving [12], math [13], and science education [14], among others. Unfortunately, there is scant research on identifying the various prerequisites for robotics-based learning. Section 3 highlights the study of [5] that focused on determining the essential prerequisites for supporting students in robotics-based STEM learning.

Along with preparing students to engage in robotics-infused learning, teachers are required to adopt instructional practices that ensure that their students are involved in learning. For example, a qualitative research study was conducted in a middle school in the southeastern US with four science teachers and 70 students [15]. The study explored the role of instruction in integrating STEM topics via robotics-based science activities. Specifically, the teachers adopted student-centered, problem- and project-based activities that allowed their students to actively engage in authentic tasks requiring problem-solving linked to real-world challenges. The instructional approaches employed by all the teachers avoided using the passive lecture method. Moreover, the students were engaged in curriculum-aligned collaborative projects that included formulating solutions, brainstorming, discussions, answering questions, etc. The teachers reported that their students were involved in the activities and acquired problem-solving skills through these experiences. Section 4 showcases the study of [6] that analyzed and recommended several instructional practices deemed successful in embedding robotics for classroom use.

To maximize the benefits of using robotics in classrooms, it is essential that students perceive it as an authentic tool for learning. Prior studies on examining student perceptions concerning their intent to pursue college majors or careers in science or technology have found that students have a favorable outlook toward robotics [16]. Yet another study conducted in Chile revealed that over 90% of participants in a robotics workshop were satisfied with learning and over 80% expressed interest in pursuing a career in engineering after the workshop [17]. The success of the workshop in [17] was attributed to participants' self-motivation since they had no prior experience in learning the topics through the robotics approach. The studies of [16] and [18] from rural Illinois and South Korea, respectively, have also showcased positive attitudes toward the incorporation of robotics in learning. As an equitable and accessible means for pursuing STEM, the study of [16] revealed that middle school female-identifying students expressed positive perceptions toward careers in robotics more than male-identifying students. The studies of [16–18] present foundational aspects of using robotics in classrooms and fostering learning environments that successfully integrate robotics in STEM learning. Nonetheless, our understanding about the factors that influence the beliefs of students concerning robotics remains nascent at best. The study of [7], described in Section 5, responds to this issue.

3. Prerequisites for robotics-based STEM lessons

Prerequisites refer to the required knowledge and skills that students must possess to effectively learn and apply a new concept [19]. Having prerequisites positively influences learning since it enables students to draw linkages between the old and new knowledge and it enhances their capacity to learn [19]. In a technology-enabled learning environment, it is critical to identify, plan for, and activate the prerequisite knowledge to enact effective lessons that make use of technology as an authentic instructional tool as opposed to a mere administrative tool. In a similar vein, integrating robotics to enhance STEM learning requires the design and development of certain activities that need to be performed before classroom lesson execution.

Since conventional STEM learning is concerned with merely science and math instruction, it does not consider the likelihood of robotics integration into the curriculum [5]. Absence of prior exposure to robotics-based activities may inhibit students, limiting their ability to adequately participate in and benefit from such lessons [5]. By identifying prerequisites and accordingly preparing students for them can relieve students' feelings of stress and anxiety, ascertain their misconceptions, gauge their interest in robotics-enabled lessons, and suggest relevant curriculum and instructional strategies to teachers [5]. Deficits in prerequisite knowledge may cause delays in learning development or lead to an incomplete understanding of the subject matter [5]. The following subsections describe a study that was conducted to identify prerequisite knowledge and skills for students to succeed in robotics-based STEM lessons.

3.a. Design of the study [5]

Twenty-three science and math teachers were recruited to participate in the study. They attended a three-week summer professional development program that enabled them to design and implement robotics-based science (e.g., displacement, energy, cell division, biological adaptation, osmosis) and math (e.g., ratio and proportion, number line, function, least common multiple, expressions and equations) lessons. The professional development and research facilitation team consisted of engineering and education graduate researchers, postdoctoral researchers, and engineering and education faculty members. During the academic year in regular class sessions, the teachers began engaging students in robotics-based science and math lessons from the workshop.

The study was conducted in two parts. First, the teachers and researchers collectively brainstormed and self-reflected to populate responses to a survey that focused on determining prerequisites for robotics-based classrooms. The survey also asked a question concerning the level of necessity of each of the prerequisites identified. Second, through classroom observations and interactions, the teachers and researchers noted the prevailing state of students' prerequisite knowledge and skills. To facilitate this, students were selected at random by the teachers and researchers and observed while participating in robotics-based learning activities. In support of this study, the teachers brought their experiences with middle school curricula, K-12 STEM standards, knowledge of their students' backgrounds, classroom environment, robotics integration, social and behavioral knowledge, etc.

Data was collected and analyzed for both parts of the study. For the surveys, a cumulative list of all the suggestions was curated and the frequencies for each of the common prerequisites was calculated. Average scores for each prerequisite were determined for the question on the level of necessity, and finally, emergent categories and themes were determined along with their relative necessities.

3.b. Outcomes of the study [5]

The analysis revealed several prerequisites that were deemed valuable for successfully integrating robotics-based lessons. Computational thinking was found to have the highest value followed by behavioral and social skills, laboratory and technical skills, engineering skills, design skills, and disciplinary content knowledge. It is important to note that students may already satisfy some of the requirements based on their current middle school STEM learning experiences. Teachers are encouraged to use and contextualize the following prerequisites to design their lessons and maximize the impact of the lessons. Details of each of the prerequisite themes can be found below.

Computational thinking ability was identified as a key requirement to succeed in robotics-based learning activities. Computational thinking encompasses cognitive functions and skills involved in the discovery of concepts and processes to solve problems [20]. It is an essential skill that can be acquired and practiced by anyone and it is not the sole domain of computer scientists. According to Wing [10], computational thinking utilizes a range of tools, concepts, and theoretical underpinnings fundamental to computer science for problem-solving, system design, and understanding human behavior. While it is the concept behind computation, it is not limited to the skill of computing or programming itself [10]. The theme of computational thinking in this study encompassed activities such as thinking and reasoning skills, creativity and imagination, systems thinking and systems modeling, anticipating consequences of results, problem-solving speed, block-based programming, decision-making, and ability to develop a hypothesis, among others [5]. Furthermore, the study revealed that participation in robotics-enabled STEM lessons can lead to an increase in students' computational thinking skills.

Behavioral and social aptitudes were identified as pivotal because collaboration and social relationships during robotics activities are key to a successful lesson execution. The activities include teamwork not only to work on a problem and subsequent investigations but also to manage resources such as robotics kits, instruments, worksheets, etc. In the study, this theme of behavioral and social qualifications included collaboration abilities, learning attitudes, ability to share an idea with others, time management, and learning aptitude.

Laboratory and technical skills of using robots are required for students to perform hands-on activities with robotics tools. Additionally, appropriate usage of laboratory instruments such as measuring tape, timer, wire gauze, other construction materials, etc., is essential otherwise students may not be able to perform robotics activities. Understanding, ability, and skills to use lab equipment and performing activities with safety were the prerequisites suggested by the teachers and integral to this theme.

Engineering prerequisites are necessary to integrate robots in classrooms. This includes vocabulary and knowledge of using parts for robot chassis, drive mechanism, sensing, actuation, and programming as well as the control, communication, power setup, and troubleshooting methods to build and use a robot as a learning tool. Thus, it is important for learners to be familiar with different robotics parts to be able to assemble them in the desired manner.

Design skills are important for robot assembly and reassembly during the lesson. Although teachers can guide learners in building basic robots for the lessons, some activities require learners to utilize additional accessories such as grippers and sensors based on different exploration activities. These scenarios require the learners to brainstorm and create robots and, thus, require design skills.

Disciplinary content knowledge of science and math is necessary for an effective implementation of the lesson. In its absence, students may become confused and unable to fully apply new knowledge gained through educational robotics. This also includes basic computer literacy in case of programming on laptop/tablet devices.

Subsequent study observations of the students on their prevailing status of prerequisite knowledge and skills revealed that they possessed *behavioral and social skills*, *laboratory and technical skills*, and robot *design* skills through their daily interactions, classroom environment, and prior experiences. *Engineering* prerequisites and *computational thinking* skills were found to be inadequate given students' lack of exposure to similar activities in the past. In certain instances, the students exhibited the knowledge of engineering terminologies and computational thinking, but it was insufficient to prepare them for robotics-based lessons. In particular, the students were found to be struggling with troubleshooting and programming the robot, reasoning the purpose behind the activities, analyzing their findings, and drawing conclusions based on the activities.

3.c. Recommendations to meet the prerequisites for robotics-enabled learning [5]

The study discussed prerequisites that the students did and did not meet for robotics-enabled STEM lessons. Following are some recommendations suggested by the researchers to plan for these requirements ahead of implementing the lessons.

First, training sessions can be offered for students to learn engineering concepts of robotics such as actuation, sensing, and control. Moreover, the use of basic laboratory instruments can be demonstrated prior to beginning robotics lessons. A training session can also be an avenue to discuss general social, behavioral, and managerial attitudes that will allow for a seamless flow of instruction. Next, students can be introduced to programs, lessons, and activities that encompass computational thinking skill development. Active, hands-on, and inquiry-based programs that allow learners to think critically, problem-solve, conduct investigations, analyze data, participate in block-based programming, etc., can enable them to participate in robotics activities.

Second, participating in robotics-based activities and receiving detailed ongoing feedback from instructors can gradually improve students' prerequisite knowledge. This method may have little impact on the lessons at the beginning because learners may lack prerequisite knowledge to optimally learn through these experiences, but it may result in overall positive learning if the lessons are scaffolded to meet learner requirements. Furthermore, the approach may not be applicable where robotics is used to teach a series of lessons, each with varied contexts, specifications, and skills demanded.

Third, if schools and institutions plan to integrate robotics for regular curriculum, then the existing curriculum should undergo changes that involve and continually generate prerequisite knowledge for robotics-enabled STEM learning. On the basis of the list of prerequisites provided in the previous subsection and through their knowledge of student backgrounds, teachers can assess learners on their current status of prerequisite knowledge to design their lesson objectives. Such an assessment is desirable since a lack of prerequisites may affect the effectiveness of robotics-based lessons.

This study elucidated the importance of understanding prerequisites required for a robotics-based STEM lesson and identified computational thinking, behavioral and social skills, laboratory and technical skills, engineering and design skills, and disciplinary content knowledge as pivotal to its success. The study also proposed some relevant and scaffolded instructional practices for educators that can help in building these knowledge and skills among their learners. These practices can aid in enhancing students' overall learning performance in addition to improving outcomes in robotics-based activities.

4. Instructional practices for effective robotics-based lessons

In constructionist learning, students assume the ownership of creating their understanding of a topic by exploring it, while working with materials and constructing artifacts, as opposed to relying on teachers as the sole knowers and providers of information in the classroom [21]. To engender a constructionist setting of learning, teachers ought to play the role of facilitators who foster inquiry, investigations, analysis, synthesis, sense-making, and reflection among students.

Moreover, teachers must shift away from the role of a 'content-expert' and instead encourage students to interact with peers and learning artifacts to acquire new knowledge. Such a learning framework encourages students to take the stance of an active participant in constructing their learning and stay engaged through such an experience. In this spirit, for promoting robotics-enabled learning, teachers must allow students to explore and examine concepts by experimenting with the robot, instead of merely using it as an object to illustrate a concept through didactic instruction. Moreover, the problems and concepts being probed by students must be framed in contexts that are deemed meaningful and relevant by them. Then, the use of robotics for solution discovery may drive along imaginative and flexible trajectories that do not necessarily lead to one 'right' answer. Yet, the setting and process of such robotics-infused constructionist learning help students prototype and test varied strategies on their own and obtain feedback from peers and teachers, all of which can further their learning.

Instructional practices refer to the teaching methods that guide students in achieving desired learning outcomes. Effective instructional practices that are aligned with the constructionist theory include promoting interactions, engendering active learning opportunities, supporting motivation for learning, and providing regular feedback [22]. Moreover, a productive enactment of constructionist learning mandates that teachers possess a deep understanding of student misconceptions and employ instructional practices that catalyze a student-initiated conceptual change through experiential investigations [23]. Finally, the following practices have been identified as efficient for the teaching of science and math [24]: learning through artifacts (e.g., manipulatives and technology), experiences (e.g., hands-on, cooperation, and discussion), and activities (inquiry, question, analysis, reflection, problem-solving, facilitation, and assessment).

Drawing from earlier research [25] on building and sustaining student attitudes in science and math, it is essential that effective instructional practices be embedded in robotics-enabled learning environments due to their promise for fostering student interest in STEM learning. Prior studies have also identified pedagogical strategies that are particularly effective for robotics-infused STEM learning such as scaffolding [26], visual modeling [27], and project-based learning [28]. Additionally, it is critical to incorporate strategies for adequate management of the classroom, for example, students, time, robots, etc., for effective incorporation of robotics-based learning [29]. Ineffective instructional practices may produce disinterest and disengagement in learning despite the presence of hands-on technology tools in the classroom. This section describes a recent study that sought to systematically identify, examine, and analyze different instructional practices that enable successful robotics-based STEM lessons.

4.a. Design of the study [6]

Twenty teachers who had implemented robotics-enabled STEM activities in their classrooms were recruited for this study. These teachers had previously been a part of a robotics-based hands-on professional development summer program that helped them in adopting best teaching practices for student learning success. During the summer program, they had engaged in experiential learning with robotics-based lessons on varied science and math topics, for example, force, displacement, number line, algebraic expressions, etc. For the classroom implementation examined in this study, one science and one math topic were collectively identified by the teachers and researchers. The teachers developed one robotics lesson and one non-robotics lesson for each topic. The lessons were selected as they: consisted of curriculum-relevant content, allowed content treatment through both robotics and non-robotics activities, provided an apt mix of science and math concepts, and would have been pedagogically difficult to teach through conventional methods. Specifically, on the selected topics of 'graphing and analyzing linear relationship' and 'wavelength-frequency relationship,' the participating teachers themselves developed lessons and corresponding assessments to prevent researcher-induced bias. The following three sources were selected to reliably collect and analyze data.

Surveys: Sixteen teachers responded to a survey to share their perspectives on student engagement in robotics and non-robotics activities. The survey probed teachers about student misconceptions, teacher approaches for addressing misconceptions through robotics and non-robotics activities, assessment methods, and student engagement. A subset of teachers (four) also responded to another survey on classroom management techniques.

Pre- and post-tests: Performance of 88 students on the pre- and post-tests was evaluated based on two lesson implementations and assessments designed, implemented, and graded by two teachers. In particular, the science and math tests included six and five multiple-choice items, respectively. The same items were used for the pre- and post-tests.

Observations: Classroom observations for robotics and non-robotics activities for selected classrooms were conducted by the researchers during the two lesson implementations.

Collected data was analyzed qualitatively by anonymizing and coding the responses and forming categories and emerging themes to reveal different factors that contribute to successful teaching practices.

4.b. Outcomes of the study [6]

The analysis of surveys, observations, and pre- and post-test data produced several themes that indicated positive learning gains with robotics *versus* non-robotics activities. The study also exemplified certain successful instructional practices that were used by the teachers to facilitate robotics-enabled learning. The practices are listed below.

Ascertaining student misconceptions: Teacher responses to surveys suggest that they recognized the promise of robotics activities in ascertaining and responding to student misconceptions. Following a careful consideration of the obstacles that students encounter in understanding abstract concepts or complex processes of science and math disciplines and the futility of traditional instructional methods in such situations, they examined the possible relevance of robotics-based learning for the same. Specifically, they treated concepts such as cell division, planetary separation in a solar system, relationship between wavelength and frequency, least common multiple, analyzing graph patterns, etc., with robotics-based lessons and activities to account for various misconceptions and to meet various learning requirements of students.

Identifying suitable topics for robotics activities: Teacher responses to surveys indicate that they deliberated on myriad factors to choose the lessons to be conducted using robotics. The first factor concerns exploring whether student learning of a lesson or topic can be aided through hands-on robotics activities. The second factor suggests analyzing the prospects for students to envision real-world applications and future benefits of robotics-integrated learning. The third factor advises examining abstract science concepts, which remain inaccessible with conventional pedagogy, for their alternative visualization with robotics-enabled experiential learning. The fourth factor advocates investigating whether various cognitive and non-cognitive objectives of a learning unit can be adequately fulfilled with the activities of an educational robotics lesson, instead of it serving as merely a tool for classroom engagement. The fifth factor proposes scrutinizing whether there is an adequate alignment for grade level and learning standards *versus* the robotics activities under consideration. Finally, the sixth factor counsels embedding opportunities for students to apply and hone their problem-solving skills through robotics-infused lessons.

Engendering opportunities for student engagement and motivation: All the teachers concurred that educational robotics activities intensified engagement in, heightened motivation for, and fostered greater understanding of the lessons among students. Several teachers reported that students with special learning needs demonstrated deeper engagement and greater motivation to perform robotics-based learning activities than the conventional learning approaches. There were considerable differences in learning with robotics *versus* non-robotics activities. First, with robotics activities, the students gained a clearer grasp of abstract concepts and they spent more time engaging with the task. Second, students who found it tough to focus during conventional pedagogy remained attentive and engaged in robotics-based activities and comprehended the concepts better. Third, by fostering collaborative learning, robotics-based lessons enabled the students to interact with one another and catalyze peer learning. Fourth, by offering tangible representations of abstract concepts, robotics-based lessons preserved student interest in learning, cultivated their curiosity, and deepened their motivation for learning. Finally, robotics-based lessons contributed toward students' behavioral development since, despite occasional mechanical, sensor, or programming failures, the students remained enthused about learning and endured their trials.

The engagement differences between robotics *versus* non-robotics activities can markedly contribute toward effective learning. Some of the approaches employed by the teachers to engender engagement in robotics-based activities are highlighted below. First, the students were divided into small teams wherein each team member was designated to serve in specific roles, for example, team leader, assembly engineer, designer, inventory manager, etc. Second, to promote the understanding of robots as essential to the learning task, the students were explicitly informed about the need and role of robots in the learning activities. Third, to ensure accountability, the students were alerted that the robotics activities contribute to their class grade so that they fully realized the responsibility of completing assigned tasks. Finally, the students were informed that they had obligations to their team members and to the teacher to accomplish the planned goals.

Measuring student progress through assessments: Assessments play a pivotal role in learning by helping to monitor student growth. Teacher responses to surveys described various assessment techniques that they employed to gauge student learning. Some of them included: entry and exit tickets, pre- and post-tests of content knowledge, online assessments, observations of student work, follow-up questions, self-assessment surveys and checklists, quality of hands-on activities, data collection and analysis, and quality of student responses on assigned worksheets. Moreover, a teacher reported adopting a rubric, which accounted for social, collaborative, and analytical skills, to gain a holistic view of student learning with robotics.

Creating classroom management structures: Teacher responses to surveys offered the following classroom management techniques that supported smooth conduct of lessons and robotics activities. The first technique suggests organizing instructional binders holding pertinent information regarding hardware, software, activities, data collection, assembly procedures, general troubleshooting ideas, etc., to reduce confusion and encourage independent problem-solving. The second technique advises providing well-designed worksheets to student teams for recording observations, tabulating data, performing calculation, etc., all of which help in streamlining learning and preventing mistakes. The third technique advocates assigning to students in each team varied roles, such as leader, designer, programmer, timekeeper, note-taker, etc., which can deepen student engagement, lead to timely completion of tasks, and help in maintaining accountability and ownership in learning. The fourth technique proposes that, with each new robotics-based lesson, student roles be switched among the members of a team so that everyone gets a chance to experience various responsibilities. The fifth technique counsels forming student teams by selecting students of varied ability levels to promote peer learning. The sixth technique recommends defining classroom norms explicitly and communicating expectations clearly so that it is feasible to keep track of materials (e.g., robotics kits, instructional binders, worksheets) and to maintain discipline.

4.c. Recommendations to plan and implement effective teaching practices for robotics-enabled learning [6]

This study uncovered myriad practices that can facilitate the effective implementation of robotics lessons and activities in classrooms. Knowledge of these practices can be valuable for teachers, curriculum developers, and even school principals in planning and scheduling for effective, engaging, and meaningful lesson delivery. They are summarized below.

It is paramount that teachers have a comprehensive understanding of typical misconceptions held by students so that they can successfully remedy those by tailoring their instruction. To gain awareness of student preconceptions, teachers must carefully analyze prior classroom performance of students, observations of student works, and their own critical reflections. Through systematic exploitation of this knowledge, teachers can utilize robotics-based activities to clarify various disciplinary concepts. For example, a teacher recognized that while the notion of cell division and its steps is deemed intangible by students, they found that a robotics lesson rendered the topic to be accessible. The use of inquiry-based learning, discussions, entry slips, or lesson design frameworks such as 5E can also help surface student misconceptions. The next two chapters in this book treat the use of Bloom's taxonomy and the 5E instructional framework to design and develop robotics-based STEM lessons.

With conventional instructional strategies often being deemed as obstacles to learning about abstract science and math concepts, the use of hands-on learning through manipulatives constitutes an effective instructional practice. Even as educational robotics can be used as a technological manipulative that supports active learning, it is not a panacea to teach all complex concepts and it must be carefully examined for its suitability and integration into lessons as a purposeful learning tool. For instance, it is known that robotics-based lessons support kinesthetic learning and lend opportunities for visual modeling. As an example, interacting kinesthetically with the gears of a robot allows learners to estimate its travel distance [30]. Similarly, visual modeling of kinetic and potential energy through a zipline robot allows learners to comprehend the abstract notions of conservation and transformation of energy [31]. Thus, lessons that can provide such opportunities are considered as being amenable to be treated using robots as tools to enhance learning.

Incorporation of educational robotics fosters a collaborative learning environment wherein learners are endowed with myriad opportunities to interact with peers to discuss situations, analyze solutions, predict outcomes, and own their learning, all while addressing an assigned challenge. As an example [30], a class of seventh graders worked in groups to learn and experience the robotics design sequence by designing, constructing, and testing a robotics ball sorter. Throughout the experiential learning process, the students collaborated in their groups on tasks such as: brainstorm solutions; conceive, sketch, and construct designs;

discuss, share, and compare ideas; perform and observe experiments; and collect and record data; among others. Effective ways to promote and sustain such learning entail creating student teams, designating team roles, encouraging peer interactions, assigning collective feedback and grades, etc. For instance, according to [6], students may be designated and rotated through roles such as group leader, design engineer, assembly engineer, and inventory manager, among others, to build their abilities and to hold them accountable to each other.

Teachers can design aligned formative assessments such as entry-exit tickets, online assessments, group observations, learning portfolio, etc., to assist in strengthening the concepts and surfacing misconceptions. As one example [6], teachers may create scoring rubrics to assess the abilities of students to collaborate, understand concepts, and perform various tasks under the robotics-based STEM lessons. Moreover, classroom management structures such as setting class norms and expectations, organizing student groups, providing binders with worksheets and troubleshooting material, etc., can help in reducing confusion and help in developing independence among learners.

The study on prerequisites for robotics-based learning [5], discussed in Section 3, also identified several relevant practices that must be carefully considered to prepare for the effective implementation of robotics-based lessons. First, it suggests ascertaining relevant situated scenarios wherein the instructional design effectively incorporates robotics as a pedagogical tool as opposed to a technocentric tool. Second, it proposes scrutinizing the lessons, robot's activities, and situations to ensure that they do not lead to misconceptions, cause safety concerns for students, or require long durations for implementation. Third, it advises designing the materials, descriptions, activity sheets, etc., for a robotics-based lesson well in advance. Fourth, it advocates predicting the impact of robotics-enabled activities on student and teacher performance. Fifth, it counsels for providing adequate professional development to teachers to impart them knowledge, skills, and confidence for implementing robotics-based lessons independently. Sixth, it recommends planning for class logistics and management including time duration, troubleshooting support for robots, arranging required hardware and software, planning student groups, etc.

The study described in this section focused on identifying and analyzing instructional practices that can engender authentic learning opportunities through robotics-infused STEM learning. The study also recommended certain practices such as: surfacing student misconceptions, carefully ascertaining topics that can be taught via robotics, creating opportunities for student engagement, measuring progress through assessments, and creating classroom management structures. These practices are key contributors in making robotics lessons effective and align with Bers' recommendations [9] for effective robotics-based learning experiences including an environment to promote building own projects, using robots as a tool for making abstract learning more tangible, identifying curriculum-aligned and suitable topics, and engendering opportunities to showcase student work.

5. Factors that influence student perceptions of utilizing robots as educational tools

Understanding student perceptions, that is, their beliefs, feelings, and thoughts, toward robotics as an instructional tool is critical when considering the common arguments in favor of robotics integration such as engendering engagement, creating excitement, and building active learning environment [8] and designing experiences that maximize the benefits of using robotics. The opinions of teachers and students form the basis of most robotics-based learning programs [32]. The study of [7], described in this section, fills the gap of limited research on factors that contribute toward building such perception. Thus, it sought to obtain a deeper understanding of student perceptions toward robotics-based learning concerning its immediate outcomes and their interest toward pursuing similar courses in their school curriculum. Moreover, the study aimed to identify various pedagogical techniques for robotics integration and their effects on student attitudes as each of the interactions affects their perceptions.

5.a. Design of the study [7]

For this study, four middle school classrooms were selected from four public schools in an urban school district in the northeastern US. The students in these classrooms were instructed by teachers who had previously attended a three-week summer professional development program on effectively integrating robots in their classrooms. Data was collected using two questionnaires, each administered at the start, or pre-test, and the end of the class, or post-test. The pretest consisted of Bradley and Lang's Self-Assessment Manikin (SAM) [33] with a pictographic measure for gauging emotions in the affective domains of pleasure, arousal, and dominance. For this study, only the pleasure and arousal scales were retained. The domains consisted of choices of positive, neutral, and negative affects, adapted from Feldman, Barrett, and Russell [34], such as high energy (alert), high positive (enthusiastic), high pleasantness (happy), low positive (relaxed), high negative (nervous), low pleasantness (sad), low negative (tired), and low energy (lethargic). Positive affects may cause an increase in student engagement.

The post-test consisted of the SAM scale and an accompanying qualitative question—"Would you like other classes in school to use robots too? Why?"—to prompt students in critically examining the role of robots as a learning tool. The post-test also consisted of an adapted version of the Test of Science Related Attitudes (TOSRA) [35], a validated instrument that measures science-related attitudes through subscales on social implications of science, normality of scientists, attitudes toward scientific inquiry, adoption of scientific attitudes, enjoyment of science lessons, leisure interest in science, and career interest in science. The enjoyment of science subscale was adapted and employed to gauge student engagement for the purposes of this study. The students were observed in classrooms and asked to fill the questionnaire without divulging any personal information.

5.b. Outcomes of the study [7]

Results on the SAM scale were mapped to positive, neutral, and negative affective states, and a change from negative and neutral states toward positive state was noticed among 7% of the participants. On the TOSRA questionnaire, the mean score was reported to be 25.39 on a scale of 0 to 40, where 0 reflects the most negative attitudes and 40 reflects the most positive ones. Student responses from the qualitative question on their thoughts on other classes using robots as an instructional tool were categorized into positive, negative, and neutral perceptions and then coded to obtain themes. Themes with high frequency of positive responses were 'enjoyment,' 'support of conceptual understanding,' 'hands-on,' 'shared experience,' and 'future prospects,' and themes with commonly occurring negative responses were 'not interesting,' 'robots should be used for specific subjects,' and 'prefer teachers.' Some classroom-specific observations were noted and are listed below.

Classroom A participated in a lesson on adaptation and natural selection that aimed to help learners experience the advantages of possessing different features that help species in their survival. The teacher used a variety of instructional tools such as the use of multimedia, formative quizzes, as well as robotics and non-robotics activities. The non-robotics activity tasked learners to act as finches and pick as much rice as possible from a bowl in a given time using instruments such as spoon, fork, knife, and chopsticks. Next, they were provided with LEGO robotics kits to construct a robot with attributes that would maximize its ability to survive in a given environment. Forty-five percent (45%) of students responded positively to the experience and 40% responded negatively on the qualitative question. Moreover, students did not immediately see a link between the activity and classroom instruction, possibly due to the open-ended inventive nature of the robotics activity. Another possible reason for this could be the non-essential add-on of the robotics activity as the non-robotics activity was adequate in explaining the concept.

Classroom B worked on a programming challenge on a simple mobile robot that was previously assembled by the students at the beginning of the school year. The task was to move the robot forward by a specific distance, turn, and return to the original location, while being able to sense obstacles. The teacher reiterated the concepts from previous classes, assigned students to groups to work on the challenge, and supported them one-on-one through the rest of the class. On the qualitative question, 71.43% of the students responded positively and 23.81% responded negatively. A large proportion of students felt that robots are not interesting, and observation notes reveal that this could be attributed to the frustration that the students experienced on not being able to make the robot perform as expected.

Classroom C used robots to understand the concept of addition and subtraction of positive and negative integers on a number line. The students observed the motion of a pre-programmed robot and identified algebraic statements related to this motion on their worksheets. Only two out of six student groups could complete the activity due to disruptions by software glitches in the pre-programmed robot. A vast majority of students did not respond to the qualitative question.

Classroom D integrated robots to visualize a relationship between acceleration and resultant shape of the distance–time curve. First, the students completed worksheets. Then, the teacher demonstrated the use of a robot with different acceleration profiles and projected the outputs from the sensor on the smart board. The students were encouraged to make connections between the displayed graphs and their worksheets and asked to revise their responses. Later, the teacher asked a student to move the robot with a certain speed and acceleration, which generated noisy data and engendered opportunities to discuss the outcome. This classroom revealed the largest increase in positive affective state as 80% of the students responded positively to the qualitative question and suggested an increase in their concept development. This could be a result of a well-planned and well-integrated robotics-based lesson that showcased an essential use of robots.

5.c. Recommendations on practices that can positively influence student perceptions [7]

Students' perceptions toward learning can positively contribute toward their motivation and engagement. Thus, it is imperative that educators work toward influencing learners such that they become interested in participating in robotics lessons. Based on learnings from this study, following are some recommendations that teachers can adopt for implementing robotics-based lessons.

Students may develop negative perceptions about using robots if they encounter technical glitches that cause frustration. To prevent these glitches, teachers must test both the hardware and software before the class and troubleshoot possible errors to avoid them from occurring during the class. The lessons must be planned such that robots are explicitly seen as an essential element to understand and visualize the concept at a deeper level. In the absence of this, students fail to recognize the connection between the concept and integration of robots as an instructional tool. Teachers may collaborate with engineers, researchers, or other educators to create lessons that can be readily used in classrooms. These lessons can also be disseminated widely such that more teachers can benefit from them. Practices suggested in previous sections such as developing prerequisite knowledge for robotics lessons, ascertaining student misconceptions, engendering opportunities for student motivation, using assessments, and creating effective classroom management structures may further help learners by reducing confusion and boosting their confidence in using robots as a learning tool. This may, in turn, positively influence their perceptions about robotics-based learning.

Investigating the factors that may affect student perceptions about using robotics as instructional tools and working on positively influencing them is important because they contribute toward excitement for and engagement in learning. This section describes some factors that shape student perspectives and suggests recommendations to teachers to prevent negative perceptions from arising. Some of the practices include testing hardware and software in advance, explicitly introducing the purpose of robots in a lesson, and collaborating with experts to design the most authentic learning experiences.

6. Conclusion

The chapter presents three studies conducted in middle school STEM classrooms using robotics as an instructional tool to deliver authentic learning experiences. The studies focus on three key attributes responsible for effective and successful robotics integration, that is, prerequisite knowledge and skills required to participate in these lessons, instructional practices that can lead to seamless integration and implementation of robotics tools, and factors that influence students' perceptions of these tools.

In line with Bers' recommendations [9], documenting findings and learnings is crucial to support deeper understanding and self-reflection among students since learning through robotics is a hands-on process. There can be multiple ways for learners to document their work such as design journals, worksheets, presentations, and reflection journals. Design journals can be created on paper or on a digital device and should capture students' design ideas, potential solutions, diagrams, codes, solution steps, challenges, peer feedback, etc. They should act as a medium to help analyze the problems as well as to iterate and revise their solutions. Similarly, worksheets can be a tool to serve as scaffolds for learners to record their data, perform analysis, and summarize findings. Authentic sharing of design ideas and solutions must be encouraged in classrooms. Peer presentations using computers, posters, or robotics devices can be a way to inspire exchange of ideas, deeper understanding, improved communication, and problem-solving skills. Finally, activities must have relevant spaces for student reflections that enable improvement in their critical thinking abilities and promote a positive mindset. Reflection journaling with scaffolded prompts such as—'*What was challenging in the design process? What was easy? What questions do you still have? What connections did you see between the previous unit and this unit?*' can enrich student learning experience and help in transferring the skills learned to other problems. Table A.3 in Appendix A presents a curated list of recommendations on the instructional practices that can support an effective and seamless integration of robotics in classroom learning.

7. Key takeaways

- Three attributes that make learning through robotics in classrooms successful include student knowledge of prerequisites, execution of effective instructional practices by the teacher, and student perceptions of robots as an instructional tool.
- Students may lack prior exposure to robotics-enabled lessons, which may limit their engagement and learning success [5]. Preparing them with appropriate prerequisites can help in addressing their stress and misconceptions, gauging their interest, and suggesting relevant curriculum and instructional strategies [5]. Absence of prerequisite knowledge may cause delays in learning or lead to the incomplete understanding of the concept [5].
- The first study described in this chapter revealed several prerequisites that are valuable in designing robotics-based lessons. Computational thinking was found to have the highest value followed by behavioral and social skills, laboratory and technical skills, engineering skills, design skills, and disciplinary content knowledge. Computational thinking [10] utilizes a range of tools, concepts, and theoretical underpinnings fundamental to computer science for problem-solving, system design, and understanding human behavior.
- Effective instructional practices include promoting student interactions, creating active learning opportunities, supporting learning motivation, and providing regular feedback [22]. For science and math instruction, some useful practices include employing manipulatives, hands-on learning, cooperative learning, discussions, inquiry-based approaches, analyzing thinking, reflecting, integrating technology, and authentic assessments [24].
- To incorporate effective instructional practices in robotics-based learning, the second study described in this chapter proposes practices such as: surfacing student misconceptions, carefully ascertaining topics that can be taught via robotics, creating opportunities for student engagement, measuring progress through assessments, and creating classroom management structures.
- Investigating student perceptions, that is, their beliefs, thoughts, and feelings toward robotics as an instructional tool, is critical [8] to engage them in learning and to maximize the benefits of such an integration.
- The third study described in the chapter revealed themes of positive student responses such as 'enjoyment,' 'support of conceptual understanding,' 'hands-on,' 'shared experience,' and 'future prospects,' and themes with negative responses such as 'not interesting,' 'robots should be used for specific subjects,' and 'prefer teachers.' Furthermore, the study suggested some practices that can help in positively influencing student perceptions and they include testing hardware and software in advance, explicitly introducing the purpose of using robots in a lesson, and collaborating with experts to design the most authentic learning experiences.

References

1. Alimisis, D., Educational robotics: Open questions and new challenges. *Themes in Science and Technology Education*, 2013.6(1): p. 63–71.
2. Eguchi, A., Theories and practices behind educational robotics for all, in *Handbook of Research on Using Educational Robotics to Facilitate Student Learning*, S. Papadakis and M. Kalogiannakis, Editors. 2021, Hershey, PA: IGI Global. p. 68–106.
3. Eguchi, A., Robotics as a learning tool for educational transformation, in *Proceedings of International Workshop Teaching Robotics, Teaching with Robotics & International Conference Robotics in Education*. 2014, Padova, Italy. p. 27–34; Available from: https://www.terecop.eu/TRTWR-RIE2014/files/00_WFr1/00_WFr1_04.pdf.
4. Williams, K., Kapila, V., and Iskander, M.G., Enriching K-12 science education using LEGOs, in *Proceedings of ASEE Annual Conference and Exposition*. 2011, Vancouver, Canada; Available from: https://peer.asee.org/17911.
5. Rahman, S.M., Chacko, S.M., Rajguru, S.B., and Kapila, V., Fundamental: Determining prerequisites for middle school students to participate in robotics-based STEM lessons: A computational thinking approach, in *Proceedings of ASEE Annual Conference and Exposition*. 2018, Salt Lake City, UT; Available from: https://peer.asee.org/30549.
6. Krishnan, V.J., Rajguru, S.B., and Kapila, V., Analyzing successful teaching practices in middle school science and math classrooms when using robotics (Fundamental), in *Proceedings of ASEE Annual Conference and Exposition*. 2019, Tampa, FL; Available from: https://peer.asee.org/32092.
7. Ghosh, S., Rajguru, S.B., and Kapila, V., Investigating classroom-related factors that influence student perceptions of LEGO robots as educational tools in middle schools (Fundamental), in *Proceedings of ASEE Annual Conference and Exposition*. 2019, Tampa, FL; Available from: https://peer.asee.org/33023.
8. Karim, M.E., Lemaignan, S., and Mondada, F. A review: Can robots reshape K-12 STEM education? in *Proceedings of IEEE International Workshop on Advanced Robotics and its Social Impacts*. 2015, Lyon, France. p. 1–8.
9. Bers, M.U., *Blocks to Robots: Learning with Technology in the Early Childhood Classroom*. 2008, New York, NY: Teachers College Press.
10. Wing, J.M., Computational thinking. *Communications of The ACM*, 2006.49(3): p. 33–35.
11. Weese, J.L. and Feldhausen, R., STEM outreach: Assessing computational thinking and problem solving, in *Proceedings of ASEE Annual Conference and Exposition*. 2017, Columbus, OH; Available from: https://peer.asee.org/28845.
12. Pane, J.F. and Wiedenbeck, S., Expanding the benefits of computational thinking to diverse populations: Graduate student consortium, in *Proceedings of IEEE Symposium on Visual Languages and Human-Centric Computing*. 2008, Herrsching am Ammersee, Germany. p. 253.
13. Braaten, B. and Perez, A., Integrating STEM and computer science in algebra: Teachers' computational thinking dispositions, in *Proceedings of ASEE Annual Conference and Exposition*. 2017, Columbus, OH; Available from: https://peer.asee.org/28559.
14. Council, N.R., *A Framework for K-12 Science Education: Practices, Crosscutting Concepts, and Core Ideas*. 2012, Washington, DC: National Academies Press.
15. Ntemngwa, C. and Oliver, S., The implementation of integrated science technology, engineering and mathematics (STEM) instruction using robotics in the middle school science classroom. *International Journal of Education in Mathematics, Science and Technology*, 2018.6(1): p. 12–40.

16. Rogers, J.J., *Middle School Student Attitudes Towards Robotics, Science and Technology*, M.S., Eastern Illinois University 2003, (Thesis).
17. Ruiz-del-Solar, J. and Avilés, R., Robotics courses for children as a motivation tool: The Chilean experience. *IEEE Transactions on Education*, 2004.47(4): p. 474–480.
18. Lee, E., Lee, Y., Kye, B., and Ko, B., Elementary and middle school teachers', students' and parents' perception of robot-aided education in Korea, in *Proceedings of ED-MEDIA– World Conference on Educational Multimedia, Hypermedia and Telecommunications*. 2008, Vienna, Austria. p. 175–183.
19. Eberly Center Teaching Excellence & Educational Innovation, *Assessing Prior Knowledge*. Available from: www.cmu.edu/teaching/designteach/teach/priorknowledge.html.
20. Ribeiro, L., Nunes, D.J., da Cruz, M.K., and de Souza Matos, E., Computational thinking: Possibilities and challenges, in *Proceedings of IEEE Workshop-School on Theoretical Computer Science*. 2013, Rio Grande, Brazil. p. 22–25.
21. Ackermann, E., Piaget's constructivism, Papert's constructionism: What's the difference? in *Proceedings of Constructivism: Uses and Perspectives in Education*. 2001, Geneva, Switzerland: Research Center in Education. p. 85–94; Available from: https://learning.media.mit.edu/content/publications/EA.Piaget%20_%20Papert.pdf.
22. Chickering, A.W. and Gamson, Z.F., Seven principles for good practice in undergraduate education. *AAHE Bulletin*, 1987.39(7): p. 3–7.
23. National Research Council, *Science Teaching Reconsidered: A Handbook*. 1997, Washington, DC: National Academies Press.
24. Zemelman, S., Daniels, H., and Hyde, A., *Best Practice: Today's Standards for Teaching and Learning in America's Schools*. 2005, Portsmouth, NH: Heinemann.
25. Eichinger, J., Successful students' perceptions of secondary school science. *School Science and Mathematics*, 1997.97(3): p. 122–131.
26. Chambers, J.M., Carbonaro, M., Rex, M., and Grove, S., Scaffolding knowledge construction through robotic technology: A middle school case study. *Electronic Journal for the Integration of Technology in Education*, 2007.6: p. 55–70.
27. Faisal, A., Kapila, V., and Iskander, M.G., Using robotics to promote learning in elementary grades, in *Proceedings of ASEE Annual Conference and Exposition*. 2012, San Antonio, TX; Available from: https://peer.asee.org/22196.
28. Carbonaro, M., Rex, M., and Chambers, J., Using LEGO robotics in a project-based learning environment. *The Interactive Multimedia Electronic Journal of Computer-Enhanced Learning*, 2004.6(1): p. 55–70.
29. Usselman, M., Ryan, M., Rosen, J.H., Koval, J., Grossman, S., Newsome, N.A., and Moreno, M.N., Robotics in the core science classroom: Benefits and challenges for curriculum development and implementation (RTP, Strand 4), in *Proceedings of ASEE Annual Conference and Exposition*. 2015, Seattle, WA; Available from: https://peer.asee.org/24686.
30. Moorhead, M., Listman, J.B., and Kapila, V. A robotics-focused instructional framework for design-based research in middle school classrooms, in *Proceedings of ASEE Annual Conference and Exposition*. 2015, Seattle, WA; Available from: https://peer.asee.org/23444.
31. Brill, A.S., Listman, J.B., and Kapila, V., Using robotics as the technological foundation for the TPACK framework in K-12 classrooms, in *Proceedings of ASEE Annual Conference and Exposition*. 2015, Seattle, WA; Available from: https://peer.asee.org/25015.

32. Altin, H. and Pedaste, M., Learning approaches to applying robotics in science education. *Journal of Baltic Science Education*, 2013.12(3): p. 365.
33. Bradley, M.M. and Lang, P.J., Measuring emotion: The self-assessment manikin and the semantic differential. *Journal of Behavior Therapy and Experimental Psychiatry*, 1994.25(1): p. 49–59.
34. Feldman Barrett, L. and Russell, J.A., Independence and bipolarity in the structure of current affect. *Journal of Personality and Social Psychology*, 1998.74(4): p. 967.
35. Fraser, B.J., Development of a test of science-related attitudes. *Science Education*, 1978.62(4): p. 509–515.

8

APPLYING COGNITIVE DOMAIN OF BLOOM'S TAXONOMY TO ROBOTICS-BASED LEARNING

1. Introduction

With the ever-growing emphasis on high-stakes academic achievement measures at the K-12 level, educational policy, curricular strategies, and instructional practices continue to produce narrow disciplinary foci. Confronted with such dynamics, science, technology, engineering, and mathematics (STEM) education faces the consistent challenge of integrating and aligning the disciplines to the everyday life experiences of students as well as situating the curriculum in authentic contexts [1]. Thus, it is not surprising that K-12 educational settings often entail the transmission of disparate facts, principles, and methods [2], as well as engagement of students in varied hands-on activities, without any overarching organizational principle to impart coherence or purpose [3]. Being burdened with unrelated facts, inflexible principles, and recipe-like methods [2], devoid of some larger meaning, creates a disconnect between school and students' lived experiences [4]. This deficit of personal relevance leads to a lack of STEM proficiency among students and their subsequent disinterest in pursuing STEM fields in the future [5]. Expectedly, exposure to such instructional approaches causes students to disregard the importance of education itself. Consequently, as students fail to recall, connect, understand, or apply concepts, principles, and skills taught at school, they suffer from a lack of growth in their cognitive repertoire.

Infusing school-based instruction with real-world contexts, which are deemed meaningful by students, as well as explicitly clarifying the structure and purpose of its content, can signal the relevance of education to students; enhance their capacity to recollect, understand, or apply the learned concepts, principles, and skills; and sharpen their higher-order thinking to synthesize new concepts [2]. As educators seek ways to engage students in a learning of the aforementioned

DOI: 10.4324/b23177-11

variety, they often face impediments in creating active learning experiences, especially when using technology as a tool to support teaching [6, 7]. Bloom's taxonomy [8] has emerged as an important tool that educators can rely upon in designing deep, active learning experiences and for promoting higher-order thinking. Since its conceptualization in the 1950s, the taxonomy has become a popular framework to serve the needs of educators in a wide array of situations. For example, it has been applied in diverse settings to develop curriculum, design instruction, and create assessments, among others, all for stimulating the growth of learners and for assisting educators globally [9].

In recent years, educational robotics has been recognized for its inherent potential in visualizing, accessing, and experiencing abstract disciplinary content owing to its ability to foster interactive, interdisciplinary, hands-on, and collaborative learning [10, 11], all of which aid in practicing and cultivating higher-order thinking. Moreover, educational robotics has the ability to motivate student engagement in STEM disciplines by making learning relevant, appealing, and accessible [12]. Thus, the incorporation of Bloom's taxonomy with educational robotics can help render authentic learning experiences with a unique focus on the development of higher-order thinking among students. In this vein, this chapter explores the role of the cognitive domain of Bloom's taxonomy in developing meaningful learning plans. Moreover, it elucidates the taxonomy's integration with robotics-based educational activities to design learning objectives, assessments, and activities of two lessons. Such an approach can help to promote higher-order understanding of abstract concepts as well as deeper engagement in learning even beyond the K-12 environment.

2. Bloom's taxonomy

The design, delivery, and assessment of effective learning experiences mandate the articulation of educational objectives that are explicit, realistic, and clear. Having such educational objectives permits a coherent framing of curriculum, streamlines the instructional practices, and conveys an unequivocal message about the knowledge and skills students will be expected to demonstrate at the conclusion of the learning episode [13]. When a learning activity entails objectives that merely stress lower-order thinking skills, such as recall or memorization, it inhibits the potential for learners' growth and their problem-solving capabilities and it causes them to disengage from learning. Thus, it is crucial to purposefully devise educational objectives that emphasize a progressive development [13] from lower- to higher-order thinking skills among students and empower them to actively participate in their learning. Bloom's taxonomy was conceived to enumerate a comprehensive classification of teaching and learning skills with the goal of supporting educators in defining educational objectives [8, 14]; providing a common language among examiners to discuss ideas and impart greater accuracy to test items [8, 14]; and serving as a tool to align the objectives with instructional activities and assessments [8, 15]. As one of the most cited educational frameworks [9, 16], Bloom's taxonomy is popular among a global community of educators and it continues to serve a broad array of educational purposes beyond merely being used for drafting educational objectives.

Bloom's taxonomy holistically accounts for a set of three models—cognitive, affective, and sensory/psychomotor—that play varied roles in learning and, thus, are used to classify objectives of learning with increasing levels of complexity [8]. Of the aforementioned three models, the cognitive domain centers around simple to complex thinking behaviors, academic abilities, knowledge, and development of intellectual skills [8]. Alternatively, the affective domain entails objectives that explicate attitudes, values, and feelings, and comprise of five categories ranging from simple to complex behaviors [8, 17]. Finally, the psychomotor domain consists of seven levels of skills related to physical movement, coordination, and sensory cues [17]. This chapter focuses on the application of the cognitive domain of the taxonomy to design effective learning activities.

2.a. Cognitive domain of Bloom's taxonomy [8]

Bloom's taxonomy orders the cognitive domain of behaviors and skills for learning hierarchically from lower (fundamental) to higher (advanced) levels [8, 17]. This organization method of Bloom allows the design and enactment of strategies to develop and accurately measure the learning progression of students through each level of behavior. As students seek to acquire higher-level behaviors and skills, they successively follow and build upon their learning from the lower levels, which enables them to gain a deeper understanding and engage in higher-order thinking. Designing and enacting curriculum, instruction, and assessment activities that explicitly account for, build upon, and support the cognitive domain of Bloom's taxonomy can permit students to hone their repertoire of basic and advanced levels of learning behaviors and skills. According to [8], these basic and advanced levels are organized under six categories, namely, *knowledge*, *comprehension*, *application*, *analysis*, *synthesis*, and *evaluation* [8, 16, 17], with progressively growing levels of difficulty.

First, beginning at the fundamental levels of taxonomy's cognitive domain, the knowledge level entails the retention and recollection ability for facts, patterns, structures, and methods, among others. In an educational robotics lesson, this may require knowing the names and functions of various hardware components and software elements. As one example, learners must demonstrate the ability to retain and recall that an infrared sensor can be used as a proximity sensor whereas a touch sensor can be used as a momentary switch. Second, the comprehension level requires that a learner gain an understanding about some information of interest and embed it within their own mental framework. This level requires a demonstration [18] and transfer [3] of one's understanding of the information under consideration instead of its mere retelling. For example, when asked to differentiate between the uses of various robotics sensors, such as infrared, light, touch, and ultrasonic, a learner must interpret the given information to classify, compare, and contrast their functioning and finally explain their differentiation correctly. Third, the application level demands one to draw upon their understanding of some information or process for addressing new situations. For instance, when challenged with a problem to create a line-follower robot, a learner must recollect how a light sensor works, understand its ability to discriminate between light and dark surfaces, and then apply it to follow a dark line on a light surface.

Fourth, progressing the drive toward the advanced levels of taxonomy's cognitive domain, the analysis level entails drawing connections between various aspects of knowledge and skill to make sense of an investigation, discover patterns, formulate claims, etc. For example, when the learners are assigned to determine the shortest (or quickest) route for a robot to traverse along an obstacle-ridden environment, they must create and execute a program that operates the robot along varied paths and collects the corresponding distance (or time) data that must be analyzed to discover the optimal solution. Fifth, the synthesis level refers to drawing meaning from disparate learning and crafting an innovative solution to a particular challenge. For example, by combining their know-how of various robotics components and skills, students must engage in further experimentation through alterations to their robotics hardware assembly or its programming code, all while drawing inferences that can aid in solving a new problem. With such a knowledge synthesis, students may redesign their robot to make it traverse faster, climb a steeper slope, or turn a sharp corner. Sixth, and finally, the evaluation level requires learners to make decisions, support views, and formulate judgments about their new learning. For example, by assessing and reflecting on their diverse explorations with robotics, learners must apply their newly gained knowledge to recommend a sensor, a coding decision, etc., for future problem-solving.

2.b. Applications of Bloom's taxonomy

Bloom's taxonomy is an intuitive, systematic, and easy-to-apply tool [9] that can assist teachers in analyzing and improving learning activities [8] and in supplanting mere memorization of facts, concepts, principles, and techniques with higher forms of thinking [19]. This classification system can be adopted and applied in diverse ways to design meaningful learning experiences including formulating desired educational objectives, aligning assessments to the educational objectives, and creating consistent learning plans, among others [8]. To elaborate, the taxonomy can serve as a powerful framework for teachers and curriculum designers to articulate learning or instructional objectives [20, 21], which are statements that stipulate the rationale behind the learning experience and streamline the knowledge and skills learners will gain from it. For a learning objective to be deemed effective, it must be realistic for the learners based on their context. Moreover, it must explicitly and concisely state observable and measurable evidence about what learners will be able to accomplish [21]. Each learning objective comprises one or more action verbs that are followed by the intended concept, knowledge, or skill to be learned. The action verbs are purposefully selected to illustrate gains in students' learning through the educational experience and aid in its measurement. Bloom's taxonomy helps in defining and classifying six explicit action verbs to craft learning objectives as per the intended cognitive process [20]. The hierarchical structure of the taxonomy suggests that the learners should first receive exposure to and cultivate the fundamental levels of the cognitive domain (*knowledge*, *comprehension*, and *application*) and then engage with the higher-order thinking of the cognitive domain (*analysis*, *synthesis*, and *evaluation*) [8, 15, 16]. That is, a learner needs to know about and comprehend a concept before applying it [18].

Backward design process for learning suggests that once learning objectives are specified, they must be aligned with assessments [3]. Assessment is an integral element of learning design since it helps in analyzing progress in student learning and it provides educators feedback on the instructional quality to perform necessary refinement [22]. The assessments are broadly divided into two categories: formative and summative. During the process of students' engagement in education, formative assessments serve as a mechanism to measure the attainment of learning by students as well as a feedback tool to allow students to take corrective actions for improving their performance. Alternatively, at the end of a learning episode, summative assessments quantify the level of attainment in learning outcomes by the students [22]. While designing assessments, it is essential to first establish the purpose of the assessment (e.g., assessment *for* learning or assessment *of* learning) and only then select an appropriate method (e.g., formative or summative) that will produce the most pertinent information [22]. Bloom's taxonomy can inform the design of assessments germane to educational environments that seek to foster through active learning students' higher-order cognitive domains [23], which support integration and application of new knowledge as opposed to mere recall of facts and information. By keeping the learner context and the purpose of assessment at the forefront, a specific level of Bloom's taxonomy needs to be selected for formulating the assessment question [24].

After establishing the learning objectives and finalizing the assessments for an educational episode, according to the backward design process, the next step is to create, structure, and enact active learning plans that are engaging and effective [3]. An engaging and effective learning plan (1) is informed by preconceptions, educational experiences, and interests of students; (2) embeds hands-on, authentic, real-world, and purposeful activities that are relevant to students and cater to varied learning styles; (3) advances student understanding and thinking to elevate it from a beginner to a mastery level; (4) allows for collaboration; and (5) gives time for feedback and promotes self-reflection among students [3, 25, 26]. Once again, Bloom's taxonomy provides a valuable framework to create, organize, and enact learning activities and events that align the learning objectives and related assessments to ensure that cognitive engagement and learning occur at specific and intended levels of the taxonomy [24]. Next, as learners begin to demonstrate mastery of concepts at the lower levels of cognitive domain, new learning activities and experiences can be designed and delivered to move students toward the higher levels of cognitive domain. In doing so, educators can examine the findings from assessments, completed by learners, to identify their learning gaps for the level of cognitive domain under consideration and make appropriate alterations to their lessons [24]. Finally, the aforementioned approach fosters the ability for metacognition among students [15], which, in turn, enables them to transfer their learning to different contexts.

2.c. Revised Bloom's taxonomy

While Bloom's original taxonomy of educational objectives [8] has been widely utilized by educators and curriculum designers, it has also received some criticisms. For example, it has been deemed hierarchical and, hence, unidimensional [15], which is not the case with student learning. Instead, learning is known to exhibit nonlinear characteristics [27] and, thus, the discrete and unidimensional nature of the taxonomy does not align with the cognitive spectrum of learning [28]. Finally, learning and skills deemed essential for success in the 21st century, such as constructivist learning and creative thinking, are absent in the original taxonomy and, therefore, slight modifications were made to its language and structure to meet these new needs [9, 15, 28].

Structurally, the unidimensional nature of the original taxonomy was modified to make it two-dimensional with the inclusion of a *Taxonomy Table* [9, 15]. Specifically, the table partitioned two intermingled attributes of the taxonomy, namely the noun attribute forming the *Knowledge* dimension and the verb attribute forming the *Cognitive Process* dimension [15]. In this way, the vertical axis of the table included four distinct knowledge domains of factual, conceptual, procedural, and metacognitive knowledge, and the horizontal axis extracted the six cognitive levels. In terms of language and terminology, the six levels of the cognitive domain were renamed to represent the corresponding verb forms, for example, the *knowledge* level was termed as *remember*. While the lower levels of cognitive domain remained unchanged, for the higher-order domain, the *synthesis* level was replaced with *evaluate* and the highest level was called *create* [15]. The new terms for the reorganized six levels are: *remember*, *understand*, *apply*, *analyze*, *evaluate*, and *create*. The resultant sub-categories contained over two dozen verbs that help inform pedagogical strategies and offer a more comprehensive framework to design learning objectives and integrate standards with the curriculum [9]. These revisions of the taxonomy allowed a wider range of educators to use it for varied purposes such as curriculum planning, instructional delivery [9, 15], and meeting the learning behaviors of the digital age [29].

3. Literature review on applications of Bloom's taxonomy in robotics

Since its inception, Bloom's taxonomy has been instrumental in guiding educators to design educational experiences that promote higher-order thinking to induce deeper comprehension. Subsequently, in recent years, educational robotics has attracted intense interest from educators owing to its myriad potential benefits, including the generation of active learning experiences and enhanced understanding of abstract STEM concepts via hands-on activities [30]. Together, Bloom's taxonomy and educational robotics offer a promising opportunity to create authentic learning experiences that go beyond the mere recall of facts, principles, and recipes [2]. Drawing from the varied successful applications of Bloom's taxonomy, it is possible to envision and design a comprehensive framework for robotics-based learning wherein the taxonomy aids in the articulation of precise learning objectives, development of relevant assessment plans, and design of corresponding lesson plans and activities, all adequately aligned with one another. In this section, we highlight two educational robotics studies that utilized Bloom's taxonomy. While the first study focused on the evaluation of student learning resulting from extracurricular robotics activities [31], the second study sought to develop a robotics programming sequence to induce higher-order thinking among students [32].

The study of [31] focused on evaluating the outcomes of a popular extracurricular robotics activity, namely the First LEGO League (FLL), for middle schoolers from eight schools over a period of two years. Specifically, the researchers aimed to gauge the impact of students' participation in robotics activities under FLL on engendering meaningful learning and developing higher-order thinking skills among them. Participation in an FLL competition embeds opportunities for meaningful learning since it requires students to collaboratively confront real-world problems and design robotics-based solutions to address them. With the aid of the six cognitive levels of revised Bloom's taxonomy, the researchers developed definitions for evaluation congruent to the context of FLL. For instance, the proposed evaluation design considered that students exhibited the *analyze/evaluate* level of the taxonomy when they made decisions following investigations, compared various options in making relevant design modifications, or identified the components responsible for a problem being examined. The evaluation data was gathered through observations, semi-structured interviews, and group interviews during which the students were asked to explain the engineering design of their robots for the annual FLL competition. The outcomes revealed that a majority of student groups demonstrated a learning progression by achieving the taxonomy levels of *understanding* and *applying*, some groups attaining the levels of *analyzing* and *evaluating*, and only one group reaching the level of *creating*. The researchers identified various factors that may play a role in constraining the students from accomplishing the higher-order thinking levels, for example, lack of time, pressure resulting from the competitive nature of FLL, and the absence of student-centered pedagogy. Student groups that had adequate time for their robotics exploration and were involved in student-centered pedagogy were the most successful.

The study of [32] focused on developing indicators of robotics programming skills assessment that employed Bloom's taxonomy for higher-order thinking levels. The researchers launched the study by performing a qualitative literature review to identify, develop, and synthesize various components of assessment informed by higher-order thinking needed for robotics coding skills. Consequently, the following three components of robot coding skills were identified: problem-solving, critical thinking, and knowledge transfer skills that are concerned with one's ability to solve problems step by step, create computer programs, and connect computing hardware to a robot, respectively. Next, for each of the three identified component skills, the researchers enumerated supporting indicators, illustrated higher-order thinking (e.g., analyze, evaluate, and create) from the revised taxonomy, and suggested a four-point Likert scale measurement for them. As one illustration, the problem-solving component was conceived as having five indicators, for example, one of these was "change the sequence of steps if the results are not met." Having created the components, indicators, and measures for robotics coding skills, a quantitative study was performed to test the reliability and validity of the skills instrument with the support of seven experts and 50 participants who had no prior exposure to robot programming and were enrolled in a robotics coding skills development training program. The findings of the study indicate acceptable reliability and internal consistency (Cronbach's alpha = 0.747). Overall, the results of this study suggest a possible strategy for using its skills instrument to design and evaluate related higher-order learning experiences. This study exemplifies an application of Bloom's taxonomy in supporting educators and evaluators even beyond K-12 contexts.

4. Integrating cognitive domain of Bloom's taxonomy with educational robotics to promote higher-order thinking

The use of robotics in teaching abstract STEM concepts lends itself to a multitude of opportunities to enhance student learning. Educational robotics helps in strengthening students' existing knowledge; provides an increased potential for developing their higher-level cognitive domains; and generates an increase in their motivation for engaging in new concepts through active learning [12, 33]. All of the aforementioned ideas aid in enhancing student engagement, retention of knowledge, and comprehension [2]. When the robotics-based lessons are integrated with Bloom's cognitive domain, the resulting learning opportunities can allow teachers to develop lessons that embed higher-order thinking to engage students in STEM learning [2]. Since the STEM topics are often viewed by

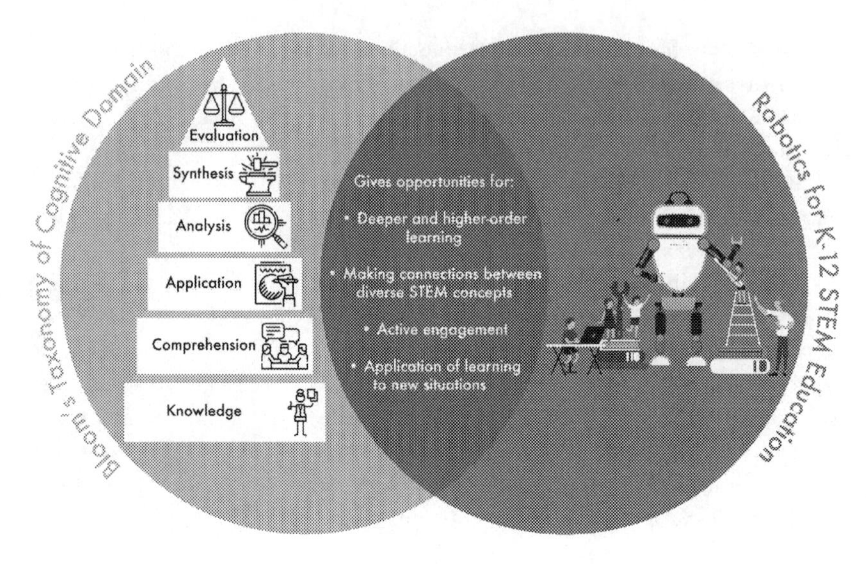

FIGURE 8.1 Intersection of the cognitive domains of Bloom's taxonomy with robotics for K-12 STEM education.

Credit line: Image designed using assets from Flaticon.com and Freepik.com

students as disconnected from their own lived experiences [4], the use of Bloom's taxonomy with educational robotics in designing an effective learning plan can help students in drawing links between varied STEM concepts, applying their learning to novel contexts, and controlling their own learning (see Figure 8.1). This section presents two robotics-based STEM lessons that employ the cognitive domain of Bloom's taxonomy to engender higher-order thinking among students. The design framework for the following lessons is inspired from the principles of backwards design [3] that entail stating objectives, drafting assessment evidence to showcase the achievement of objectives, and designing a plan to meet the objectives. See Table A.4 in Appendix A for a lesson planning template that employs Bloom's taxonomy.

4.a. A robotics-based lesson on biomimicry and echolocation using Bloom's taxonomy

The lesson plan in Table 8.1 models an application of the cognitive domains of Bloom's taxonomy in a robotics-based STEM lesson on biomimicry for an elementary school classroom [2]. Biomimicry refers to an imitation of some system from nature, such as the functioning or activities of an animal or a plant, that can be used as an inspiration to solve complex real-world problems. The lesson utilizes the idea of echolocation in bats, that is, the bat's way of detecting obstacles around it through the reflection of sound, as a model of biomimicry. The

TABLE 8.1 A robotics-based STEM lesson for elementary school

Objective: Students will be able to *evaluate* a model of echolocation in animals through biomimicry using LEGO robots.
Key questions: How do animals use echolocation for their needs?
Assessment evidence: Students will *justify* how the robotics system models biomimicry and *describe* how animals use echolocation.
Material required: LEGO robots with ultrasonic sensors attached, measuring tape, timer, graphing paper, laptops with LEGO program.

	Activities and lesson plan:
Opening	Give the students some examples and situations of biomimicry and ask them to share some examples from their prior knowledge.
Knowledge	Discuss an example and address any misconceptions about biomimicry. Ask the students to imagine a situation in which they shout in a canyon and time the return of corresponding echo to find their distance from the canyon wall. A bat's echolocation is similar wherein it emits a high frequency sound whose echo returns to the bat's ears after rebounding from an obstacle in its path. The time of rebound determines the bat's distance from the obstacle.
Comprehension	Engage the students to conduct object detection experiments using a LEGO ultrasonic sensor. Ask them to observe the functionalities of the sensor as it detects the environment around itself and displays the corresponding data. By experimenting, they understand how the ultrasonic sensor can represent the model of bat's echolocation and determine distance from the obstacle. Using their understanding of applications of the ultrasonic sensor, they perform data collection activity on three objects of their choice. Through discussions and experimentation, they gain knowledge about the sensor's interaction with respect to variations in the environment (e.g., shape/size of an obstacle) and accuracy of measuring distances.
Application	Guide the students to manipulate the threshold values for object detection by programming a robot instrumented with an ultrasonic sensor. The activity enables them to gain an insight into a robot's reaction when it encounters an obstacle. In student groups, task the students to develop their own programs with different threshold values for the ultrasonic sensor (10 cm, 25 cm, . . ., 90 cm) and to conduct investigations for assessing performance accuracy. Engage them in discussing how the model represents biomimicry, what is the role of an ultrasonic sensor, and how can the robot be made to respond to stimulus faster.

(*Continued*)

TABLE 8.1 (Continued)

Analysis	Ask different student groups to place an artifact at 100 cm and then at 200 cm distance from their robots to identify the obstacle detection threshold value for different robots. Then the student groups record and share the time it took their robots to react to the object. Each group compares their results with the other groups and draws inferences on the effect of threshold values on the robot response time.
Synthesis	Support the students to use the data collected from earlier investigations to redesign and rebuild their robot, with the ultrasonic sensor placed on the robot's front, middle, or rear, for enhancing its maneuverability. For different placements, the students determine the response time for objects placed 100 cm and 200 cm away from the robot. They analyze the results and share them with other groups.
Evaluation	Discuss students' overall synthesis of all the activities and instigate them to come up with two to three recommendations to engage in the activity more deeply next time. Encourage them to justify their recommendations based on what was learned in the class. Through a presentation, students explain how the system modeled biomimicry and how animals use echolocation.

understanding of biomimicry-based echolocation and obstacle avoidance can be transferred to various real-life situations such as a thoughtful design of autonomous vehicles or deep-sea underwater navigation. To model the aforementioned scenario, a robot instrumented with an ultrasonic sensor for obstacle avoidance is used in the lesson.

The congruence between the learning objective, the assessment, and the lesson plan is evident and caters to Bloom's taxonomy's fundamental and advanced cognitive domains. The lesson plan moves in a scaffolded manner, allowing the learners to first understand the foundational knowledge and eventually build their higher-level thinking. The lesson starts by surfacing learners' misconceptions and then develops their *knowledge* by helping them visualize the correspondence between ultrasonic sensing and echolocation by a bat. Next, through experimentation, the students deepen their *comprehension* as they explore interactions of the ultrasonic sensor with obstacles in its vicinity and their effects on the resulting measurements. Advancing toward higher levels of the cognitive domain, namely *application*, they draw on their understanding to build a robot instrumented with the ultrasonic sensor to traverse its surroundings. Next, they sharpen their *analysis* skills through experimental exploration of the robot's reactive behavior in response to the ultrasonic sensor's sensitivity settings. Having experienced the scaffolded levels of experimentation, they hone their skills of *synthesis* by interpreting findings using the data collected and enhance obstacle avoidance behavior of the robot by examining various sensor placement strategies. Finally, to grow their *evaluation* skills, the students perform experiments using the robot, obtain qualitative data, and make recommendations to enhance the design.

This lesson was implemented in an elementary school classroom. Modest gains in student learning and engagement were obtained through pre- and post-tests on the lesson concepts. A majority of the students demonstrated an understanding of the main concept of the lesson, that is, modifying the sensor placement influences the robot's interaction with and movement in its environment. A key recommendation of the study was to devote an adequate amount of time, around three to five class periods, on the knowledge and comprehension domains so that students have ample opportunities to deepen their understanding prior to moving to higher levels of the cognitive domain [2].

4.b. A robotics-based lesson on center of mass using Bloom's taxonomy

The following lesson plan in Table 8.2 illustrates an application of Bloom's taxonomy for a middle school robotics-based lesson on center of mass [11, 34, 35]. The concept of center of mass is abstract in nature and, thus, can be challenging to conceptualize, especially when considering its application in real-world situations. The lesson introduces students to a problem scenario in a hands-on manner using robotics as a tool. Moreover, it incorporates the cognitive levels of Bloom's taxonomy to facilitate a step-by-step exposure to higher-order thinking. The students play the role of an investigator to solve the problem of a truck toppling while carrying a load up or down on an incline or an automobile tipping over when making a sharp turn.

This lesson exemplifies the integration of Bloom's taxonomy in designing learning objectives, assessments, and a lesson plan that facilitate the progression from the lower to higher levels of thinking and comprehension for an abstract STEM topic. First, to develop student *knowledge*, the lesson allows them to explore the concept of center of mass by defining what it means and providing illustrative examples of its real-world uses. Having acquired an initial knowledge of the concept, the students receive an introduction to a problem scenario that they will solve as a part of this lesson. Specifically, the problem entails identifying the center of mass of a system to prevent a truck from toppling when it goes up an incline. The situation is analogous to other everyday problems involving automobile accidents that occur when a speeding vehicle attempts a sharp turn or a vehicle topples going up on a slope, etc. Second, with the initial concept and situation described earlier, the students develop an *understanding* of the topic in relation to the scenario and explain the connection. Third, to cultivate the next level of cognition, they

TABLE 8.2 A robotics-based STEM lesson for middle school

Objective: Students will be able to *illustrate* the role of center of mass of a system in a real-world scenario by conducting investigations and applying problem-solving strategies using LEGO robots.
Key questions: How can we locate the center of mass of a system?
Assessment evidence: Students *explain* the role of center of mass in a real-world scenario and *justify* a stable configuration that will prevent any crashes.
Material required: LEGO robot with a programmable brick, measuring tape, cardboard, calculator, laptops with LEGO program.

	Activities and lesson plan:
Opening	Share with students various examples to illustrate the location where the mass of an object is concentrated. Ask questions to elicit their prior understanding.
Knowledge	Showcase a video on the gravity and center of mass along with some scenarios of vehicles tipping over when making sharp turns. Introduce a real-life scenario based on events of a trucking company. The students are asked to play the role of an investigator who is seeking to determine the cause of crashes involving trucks transporting materials up on an incline. To conduct their inquiry, they are provided with a LEGO robot with a programmable brick to simulate a loaded truck and conduct experiments with it.
Comprehension	Encourage the students to draw relationships between the concept of center of mass and the truck scenario. Discuss some causes of why the truck keeps toppling. Ask them to explain the aforementioned relationship and predict the causes of crashes.
Application	Guide the students to test their predictions using different horizontal placements of the programmable brick and driving the robot up or down the incline (see Figure 8.2(a) and Figures 8.3 (a)–(b)). Alternatively, ask them to explore different vertical placements of the programmable brick and command the robot to make a sharp turn (see Figure 8.2(b)). The students record the data corresponding to various positions of the brick and the resulting stability behavior of the robot and share their responses with the other groups.
Analysis	Support the students to revisit the scenario based on their own observations and learning from the findings of their peers from other groups. Task them to design a model that prevents the robot from toppling when it goes up the incline. Monitor them as they devise the best case solution to prevent future crashes. Ask them to showcase their newly proposed designs to other groups.
Synthesis	Elicit changes in student learning and knowledge by asking them to identify the most and least stable models. Probe them to explain the rationale behind their responses. Explain that the truck with mass placed in the front will drive up an incline without toppling, and thus, the model with the brick right above the front wheels will be the most stable.
Evaluation	Support the students to justify why the robot moved differently with varying configurations of the brick and summarize the best method they would suggest to the trucking company to prevent any future crashes.

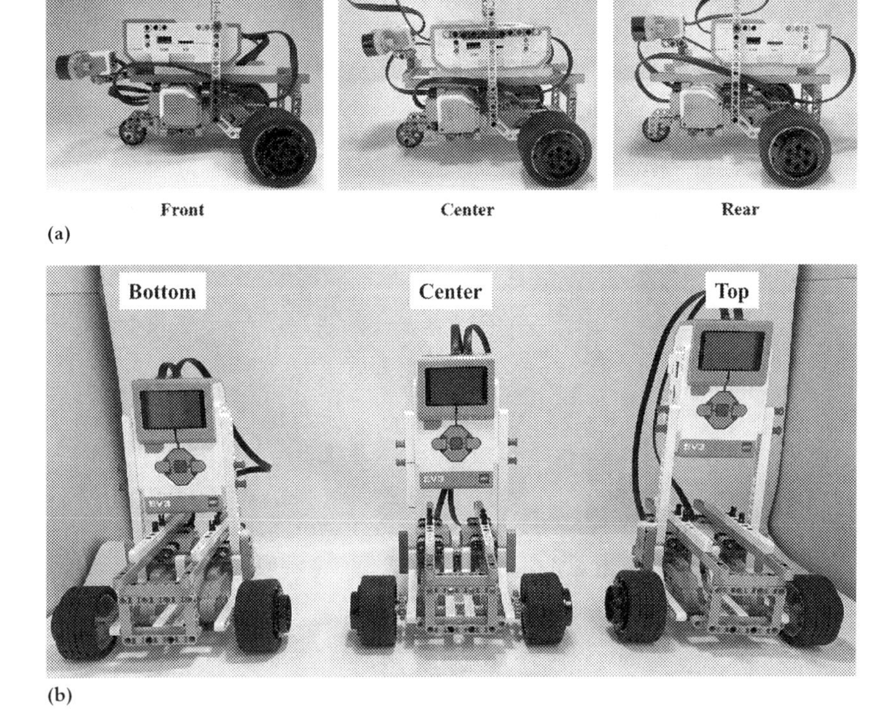

FIGURES 8.2 (a)–(b) LEGO Mindstorms EV3 brick placed on the robot structure at three horizontal locations (a) and three vertical locations (b).

apply their understanding of center of mass in conducting experiments using the LEGO robot. Specifically, they mount the programmable brick at different locations on the LEGO robot, run the robot on an incline, record the outcomes of their experiments, and then seek to identify the causes of any crashes. The student groups share their findings with their peers from other groups to gain a collective and deeper insight into the problem scenario. Fourth, in the *analysis* stage, they get an opportunity to revisit the scenario and redesign their robotics model for preventing any future crashes. With the support of data collected, they are encouraged to *synthesize* new findings about the rationale behind the placement of the brick for producing stable behavior by the robot driving up on the incline. Sixth, and finally in the *evaluate* stage, based on the argument of center of mass, the students are tasked to determine a justification for the success of their design and present a recommendation to the trucking company to prevent future crashes.

Participation in a lesson of this nature can render opportunities for students to partake in higher-order thinking and generate rich discussions. The lesson is appropriately paced with multiple opportunities that allow students adequate time to intellectually engage with the content while investigating the problem

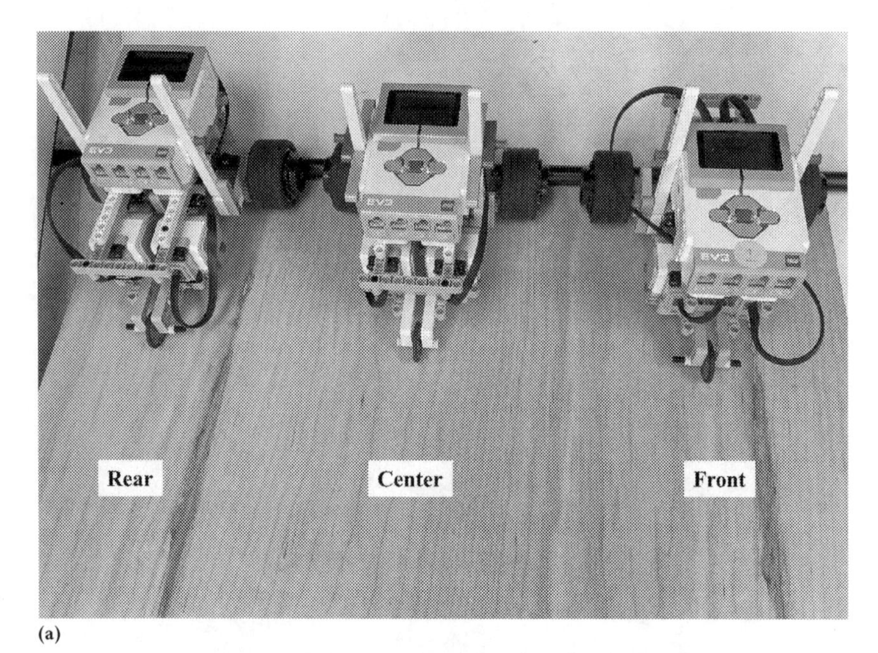

(a)

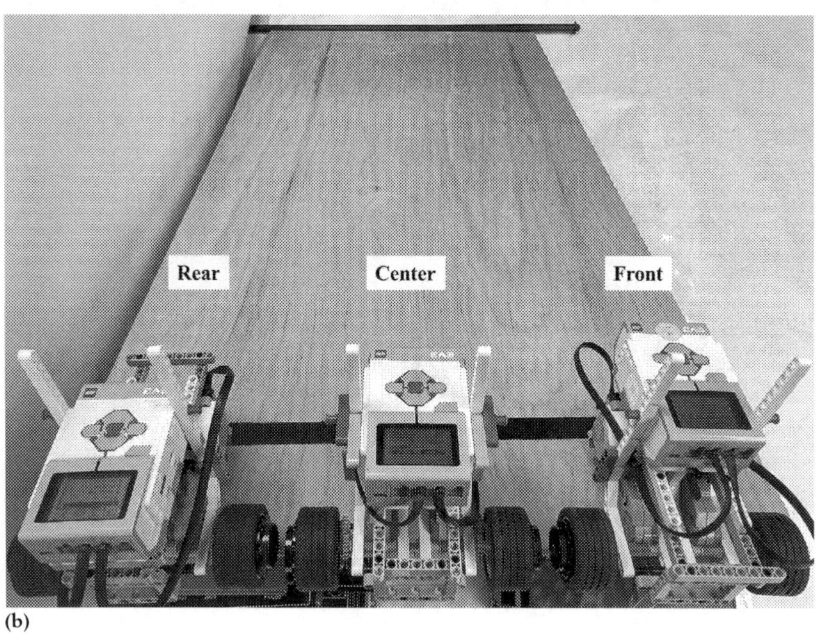

(b)

FIGURES 8.3 (a)–(b) Three robot structures with LEGO Mindstorms EV3 brick placed horizontally at three locations moving up (a) and down (b) on an incline.

and to reflect on their learning while seeking and evaluating solutions. Furthermore, myriad opportunities to collaborate and share their learning allow students to become resources for one another. Finally, physical interaction with learning material can support students in deeper comprehension of the concepts and forming a long-term association with them [2]. Overall, such an approach holds the promise of enhancing learning experiences, motivating students to engage with complex topics, and moving their status from novice to advanced learners.

5. Conclusion

Over the years, Bloom's taxonomy has gained popularity not only among K-12 teachers, but also among educators in higher education institutions. It provides a common language to educators and offers a guiding framework to design, create, enact, and assess engaging and productive teaching and learning experiences. It is a powerful tool to transform learning experiences by making them learner-centered, active, meaningful, and engaging. Bloom's taxonomy has found myriad applications in the field of education such as articulating learning objectives, devising assessments and evaluation tools, designing lesson plans, etc.

An application of the cognitive domains of Bloom's taxonomy in learning design can engender active engagement, in turn, promoting deeper and higher-order thinking among students. Integration of robotics can further aid in providing hands-on activities that contribute to increase in student motivation for learning abstract STEM concepts. The chapter presented two robotics-based STEM lessons that integrate the cognitive domain of taxonomy at several levels and that afford students the opportunity to construct, apply, and synthesize their learning. Although Bloom's taxonomy has been one of the most referred educational tools, its integration with robotics-based learning remains relatively nascent and requires further and deeper exploration. Specifically, to demonstrate the positive impact such an integration can have on K-12 STEM teaching and learning, future studies may consider long-term retention of ideas and application of newly formed understanding in other subject areas.

6. Key takeaways

- Bloom's taxonomy consists of a set of three distinct models—cognitive, affective, and sensory/psychomotor domains—that are used to classify educational objectives as well as organize assessments and learning plans into increasing levels of complexity. This promotes higher-order thinking and deeper comprehension of concepts among students.
- The cognitive domains of the taxonomy focus on knowledge and development of intellectual skills and consists of six categories—knowledge, comprehension, application, analysis, synthesis, and evaluation. Bloom's taxonomy

was revised, and the new six levels of cognitive domains include—remember, understand, apply, analyze, evaluate, and create.

- Bloom's taxonomy can be applied in educational contexts in myriad ways, including, formulating learning objectives, aligning assessments to the learning objectives, and creating congruent learning plans, among others.
- Use of robotics in STEM education strengthens students' existing knowledge, increases their chances of developing higher-order cognitive domains, and enhances their motivation for engaging in new concepts through active learning.
- Integration of robotics-based lessons with Bloom's cognitive domains can allow students to develop higher-order thinking, draw connections between diverse STEM concepts, and apply their learning to new situations.

References

1. Rennie, L., Venville, G., and Wallace, J., Making STEM curriculum useful, relevant, and motivating for students, in *STEM Education in the Junior Secondary: The State of Play*. 2018, Singapore: Springer. p. 91–109.
2. Muldoon, J., Phamduy, P.T., Le Grand, R., Kapila, V., and Iskander, M.G., Connecting cognitive domains of Bloom's taxonomy and robotics to promote learning in K-12 environment, in *Proceedings of ASEE Annual Conference and Exposition*. 2013, Atlanta, GA; Available from: https://peer.asee.org/19343.
3. McTighe, J. and Wiggins, G., *The Understanding by Design Handbook*. 1999, Alexandria, VA: Association for Supervision and Curriculum Development.
4. Kozoll, R.H. and Osborne, M.D., Finding meaning in science: Lifeworld, identity, and self. *Science Education*, 2004.88(2): p. 157–181.
5. President's Council of Advisors on Science and Technology (PCAST), *K-12 Education in Science, Technology, Engineering, and Math (STEM) For America's Future*. 2010; Available from: https://nsf.gov/attachments/117803/public/2a–Prepare_and_Inspire–PCAST.pdf.
6. Chen, F.H., Looi, C.K., and Chen, W., Integrating technology in the classroom: A visual conceptualization of teachers' knowledge, goals and beliefs. *Journal of Computer Assisted Learning*, 2009.25(5): p. 470–488.
7. Ottenbreit-Leftwich, A., Liao, J.Y.-C., Sadik, O., and Ertmer, P., Evolution of teachers' technology integration knowledge, beliefs, and practices: How can we support beginning teachers use of technology? *Journal of Research on Technology in Education*, 2018.50(4): p. 282–304.
8. Bloom, B.S., Englehart, M.D., Furst, E.J., Hill, W.H., and Krathwhol, D.R., *Taxonomy of Educational Objectives: The Classification of Educational Goals. Handbook I: Cognitive Domain*. 1956, London, UK: Longmans, Green, and Co.
9. Forehand, M., Bloom's taxonomy, in *Emerging Perspectives of Learning, Teaching, and Technology*, M. Orey, Editor. 2010, North Charleston, SC: CreateSpace. p. 41–47.
10. Brill, A.S., Listman, J.B., and Kapila, V., Using robotics as the technological foundation for the TPACK framework in K-12 classrooms, in *Proceedings of ASEE Annual Conference and Exposition*. 2015, Seattle, WA; Available from: https://peer.asee.org/25015.

11. Brill, A.S., Elliott, C.H., Listman, J.B., Milne, C.E., and Kapila, V., Middle school teachers' evolution of TPACK understanding through professional development, in *Proceedings of ASEE Annual Conference and Exposition*. 2016, New Orleans, LA; Available from: https://peer.asee.org/25720.
12. Eguchi, A., Educational robotics theories and practice: Tips for how to do it right, in *Robots in K-12 Education: A New Technology for Learning*, B.S. Barker, Editor. 2012, Hershey, PA: IGI Global.
13. Mitchell, K.M. and Manzo, W.R., The purpose and perception of learning objectives. *Journal of Political Science Education*, 2018.14(4): p. 456–472.
14. Simpson, E.J., *The Classification of Educational Objectives, Psychomotor Domain*. 1966, Urbana, IL: University of Illinois.
15. Krathwohl, D.R., A revision of Bloom's taxonomy: An overview. *Theory Into Practice*, 2002.41(4): p. 212–218.
16. Seddon, G.M., The properties of Bloom's taxonomy of educational objectives for the cognitive domain. *Review of Educational Research*, 1978.48(2): p. 303–323.
17. Binugroho, E.H., Ningrum, E.S., Basuki, D.K., and Besari, A.R.A., Design of curriculum matrix for robotics education derived from Bloom's taxonomy and educational curriculum *IPTEK Journal of Proceedings Series*, 2014.1(1).
18. Adams, N.E., Bloom's taxonomy of cognitive learning objectives. *Journal of Medical Library Association*, 2015.103(3): p. 152–3.
19. Churches, A., *Bloom's Digital Taxonomy*. 2008; Available from: http://burtonslifelearning.pbworks.com/f/BloomDigitalTaxonomy2001.pdf.
20. Anderson, L.W., Krathwohl, D.R., Airasian, P.W., Cruikshank, K.A., Mayer, R.E., Pintrich, P.R., Raths, J., and Wittrock, M.C., *A Taxonomy for Learning, Teaching, and Assessing: A Revision of Bloom's Taxonomy of Educational Objectives (Complete Edition)*. 2001, New York: Longman.
21. Chatterjee, D. and Corral, J., How to write well-defined learning objectives. *The Journal of Education in Perioperative Medicine: JEPM*, 2017.19(4): p. 1–4.
22. Fisher, M.R., *Student Assessment in Teaching and Learning*. Center for Teaching, Vanderbilt University; Available from: https://cft.vanderbilt.edu/student-assessment-in-teaching-and-learning/.
23. Tabrizi, S. and Rideout, G., Active learning: Using Bloom's taxonomy to support critical pedagogy. *International Journal for Cross-Disciplinary Subjects in Education (IJCDSE)*, 2017.8(3): p. 3202–3209.
24. Preville, P., *The Professor's Guide to Using Bloom's Taxonomy: How to Put America's Most Influential Pedagogical Model to Work in Your College Classroom*. 2019; Available from: https://ivylearn.ivytech.edu/courses/1094008/files/94634203.
25. Milkova, S., *Strategies for Effective Lesson Planning*. 2012; Available from: https://crlt.umich.edu/gsis/p2_5.
26. Baird, J.R., Fensham, P.J., Gunstone, R.F., and White, R.T., The importance of reflection in improving science teaching and learning. *Journal of Research in Science Teaching*, 1991.28(2): p. 163–182.
27. Berger, R., *Here's What's Wrong with Bloom's Taxonomy: A Deeper Learning Perspective*. 2018, Education Week; Available from: www.edweek.org/education/opinion-heres-whats-wrong-with-blooms-taxonomy-a-deeper-learning-perspective/2018/03.
28. Tutkun, O., Güzel Candan, D., Koroğlu, M., and İlhan, H., Bloom's revised taxonomy and critics on it. *The Online Journal of Counselling and Education*, 2012.1(3): p. 23–30.
29. Churches, A., Bloom's taxonomy blooms digitally. *Tech & Learning*, 2008.1: p. 1–6.

30. Eguchi, A., Robotics as a learning tool for educational transformation, in *Proceedings of International Workshop Teaching Robotics, Teaching with Robotics and International Conference Robotics in Education*. 2014. Padova, Italy. p. 27–34; Available from: https://www.terecop.eu/TRTWR-RIE2014/files/00_WFr1/00_WFr1_04.pdf.
31. Kaloti-Hallak, F., Armoni, M., and Ben-Ari, M., The effect of robotics activities on learning the engineering design process. *Informatics in Education*, 2019.18(1): p. 105–129.
32. Lertyosbordin, C., Maneewan, S., and Easter, M., Components and indicators of the robot programming skill assessment based on higher order thinking. *Applied System Innovation*, 2022.5(3): p. 47.
33. Eguchi, A., Theories and practices behind educational robotics for all, in *Handbook of Research on Using Educational Robotics to Facilitate Student Learning*. 2021, Hershey, PA: IGI Global. p. 68–106.
34. Gagliardi, I. and Ramirez, D., *Center of Mass*. 2015; Available from: http://mechatronics.engineering.nyu.edu/pdf/center-of-mass.pdf.
35. Moorhead, M., Elliott, C.H., Listman, J.B., Milne, C.E., and Kapila, V., Professional development through situated learning techniques adapted with design-based research, in *Proceedings of ASEE Annual Conference and Exposition*. 2016, New Orleans, LA; Available from: https://peer.asee.org/25967.

9

USING THE 5E INSTRUCTIONAL MODEL TO DEVELOP ROBOTICS-BASED SCIENCE UNITS

1. Introduction

Traditional instructional approaches can cause learners to disengage from science, technology, engineering, and mathematics (STEM) disciplines as they do not foster a deep understanding of concepts and fail to establish their implications to real-life situations [1]. Failure of classroom STEM instruction to fully engage students, beyond the mere consumption of prevailing disciplinary orthodoxy, can cause a decline in student motivation to pursue these fields [2]. Inquiry-based learning constitutes a promising framework under which learners are encouraged to formulate questions, examine scenarios, and produce explanations through active participation in addressing problems from real-world contexts. Such an approach to learning can aid students in gaining an improved understanding of ideas, retention of concepts, and perceptions toward science [3]. Despite recognizing the affordances of inquiry-based method, educators often deem it challenging to design open-ended exploratory learning experiences [4].

One inquiry-based model for promoting collaborative and active learning, which was developed in 1987 by the Biological Sciences Curriculum Study, has come to be known as the 5E instructional framework [5]. Grounded in the constructivist approach to learning [6], the 5E instructional model includes five learning phases—*engage, explore, explain, elaborate*, and *evaluate* [5]. In a 5E learning environment, to solve problems and investigate new ideas, students ask questions, observe phenomena, analyze data, synthesize findings, and present conclusions. In its role as a curriculum and instruction planning approach, the 5E model can guide educators to methodically design inquiry-based lessons. Broad adoption of the 5E model can contribute to bridging the prevailing STEM engagement gap

DOI: 10.4324/b23177-12

by making student learning more concrete and enduring [7], improving decision-making [8], and enhancing attitudes toward learning [9, 10].

Technology-aided interventions can support learners in improving their scientific exploration and problem-solving skills [11, 12]. As a technological tool, robotics creates accessible and engaging experiences by making the abstract concepts easier to comprehend [13]. Thus, the integration of inquiry-based pedagogy with robotics-aided learning can lead to active and hands-on experiences. Use of robotics in science classrooms affords additional opportunities for conceptual change as students interact with varied engineering tools, apply their learning to solve problems using a robot as a tool, and grow their interest in STEM disciplines [14]. This chapter employs the aforementioned rationale for integrating the 5E instructional model to engender active engagement with the use of robotics as a tool for science learning. First, the chapter describes the 5E framework with its phases and effectiveness. Second, it provides a brief review of several prior studies that have used the 5E model with high-fidelity for robotics-based learning. Third, through two exemplar learning units, the chapter showcases the efficacy of the 5E model in designing standards-aligned robotics-enhanced explorations, including their ability to promote hands-on engagement and make abstract concepts easier to understand. Finally, the chapter offers recommendations for educators to effectively employ the 5E model in designing robotics-based learning experiences for students and teachers.

2. The 5E instructional model

Many educators in the early 20th century considered that a collection of facts formed the discipline of science, which was best taught through didactic instruction and whose learning required rote memorization [4]. Beginning in the middle of the 20th century, science education underwent a paradigm shift with inquiry-based pedagogy emerging as a prominent strategy [4], which received the recommendation by Project 2061: Science for All Americans [15] and was incorporated in the National Science Education Standards by the National Research Council [16].

Inquiry-based learning is a technique that allows students to formulate or discover concepts by asking questions; conducting observations; investigating phenomena; performing comparisons; making predictions; using tools to collect, explore, organize, and interpret data; drawing conclusions and framing explanations; and communicating results [8, 16]. It has been noted that inquiry-based science curriculum leads to active participation [3], eliminates the focus from rote memorization [17], and is effective in learning for understanding [3].

Despite appreciating the need for inquiry learning, educators often find it challenging to design and embed learning experiences that foster active, participatory investigations in their classrooms [4]. One strategy to incorporate inquiry in lesson plans is the use of a learning cycle approach [4, 5], which facilitates constructivist learning and is consistent with Jean Piaget's developmental theory

[6]. The 5E framework has been adapted from the Atkin–Karplus learning cycle (also referred to as the Science Curriculum Improvement Study cycle) [5, 17]. The three-phases of Atkin-Karplus' learning cycle approach—exploration, concept introduction, and concept application—focus on first allowing the students to explore a new topic, then getting formally introduced to the new topic, and finally applying the topic to varied situations (see Table 9.1) [5].

Building on the three phases of Atkin–Karplus' learning cycle model, the 5E instructional model integrates cooperative learning [18] and includes phases where students get sufficient opportunities to: examine their prevailing views of knowledge, reformulate and update their ideas, and assess their learning [19]. Thus, the three phases of the learning cycle—*exploration, concept introduction*, and *concept application*—align, respectively, with the explore, explain, and elaborate phases of the 5E model. Engage and evaluate were introduced as the new phases in the 5E model (see Table 9.1), making it a total of five phases that each start with the letter E–*Engage, Explore, Explain, Elaborate*, and *Evaluate*. As intended by the researchers, the 5E model is: grounded in research, constructivist in nature, coherent in instructional design, and straightforward in its utility [19].

2.a. Description of the 5E instructional model

As observed earlier, with the constructivist approach, the 5E model fosters learning through inquiry to catalyze a conceptual change in learners [20]. To facilitate learning through this model, students must: be cognizant of their prior knowledge or misconceptions, be open to new structures of knowledge, integrate the new concepts into their current conceptual framework, and build their own learning [20]. According to Bybee et al. [5], through the five phases of the model learners gain clarity for the science concepts. The instructional model goes through the five phases as illustrated in Figure 9.1.

TABLE 9.1 Alignment between Atkin-Karplus Learning Cycle and 5E Instructional Model

Atkin–Karplus Learning Cycle	*Summary [5]*	*Alignment with 5E Model*
		Engage (new)
Exploration	First experience with the concept.	Explore
Concept introduction	Introduction to new terms and concepts.	Explain
Concept application	Application of concepts in related but new situations.	Elaborate
		Evaluate (new)

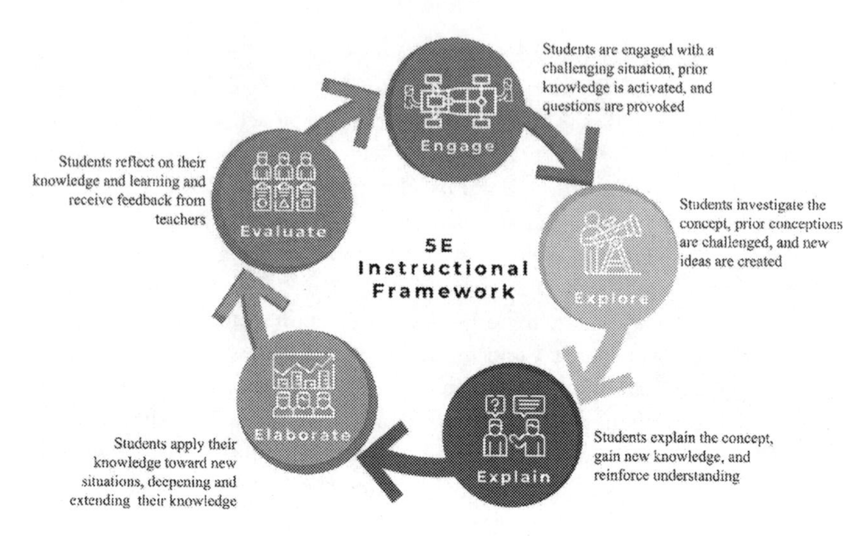

FIGURE 9.1 The 5E instructional model aligned with learner behaviors.

Credit line: Image designed using assets from Flaticon.com

The *engage* phase focuses on accessing and gauging learners' prior knowledge. Actively engaging learners in a concept, which is new to them, can help in activating their prior conception of knowledge, drawing links between prior understanding and current learning tasks, and engendering curiosity. Thus, questions and activities are intentionally designed to trigger, reveal, and connect with learners' prior understanding and experiences, arouse their interest in the new topic, initiate learning, and structure their thinking in support of the desired learning goals [5].

The *explore* phase supports learners to creatively employ their prior knowledge to examine scenarios, test hypotheses, make connections with new material, conduct experimental investigations, predict the outcome of a situation, try alternative approaches, propose new ideas, and thus produce a conceptual change [5].

The *explain* phase embeds opportunities for learners to construct and explain their understanding of concepts gained through the activities of the previous two phases. This phase encourages teachers to introduce and explain new concepts, skills, and behaviors that can transition students toward a deeper learning [5].

The *elaborate* phase challenges conceptual understanding of learners through application-based activities and experiences to strengthen and broaden their reasoning. Learners draw upon their prior understanding, apply newly acquired knowledge to varied situations, and seek evidence to draw conclusions. Through these experiences, students take a deeper dive and make a broader foray in understanding the concept and gain additional higher-order thinking skills to apply their learning to varied situations [5].

The *evaluate* phase embeds opportunities for students to summarize, assess, and demonstrate the mastery of their learning, skills, and abilities. Furthermore, it allows teachers to examine learners' knowledge comprehension, process skills, and behaviors—all of which contribute to their overall progress toward lesson goals [5].

For further elaboration, Table A.5 in Appendix A explains the five phases with respect to ideal behaviors of learners and corresponding instructional strategies. It also clarifies learner and teacher actions that are inconsistent with particular phases and should be avoided in practice. For instance, the engage phase focuses on surfacing learners' prior knowledge conceptions and hooking them in learning. Hence, learners should engage in activities that spark their curiosity and showcase their current knowledge, and teachers must act as facilitators to elicit these responses. In this phase, learners should not seek a single solution, and teachers should refrain from answering questions or explaining the concept. Instead, teachers must direct their efforts to heighten the curiosity of learners and support them in uncovering connections.

2.b. Effectiveness of the 5E instructional model

Since its development, the 5E model has found varied applications in designing inquiry-based learning experiences in science settings and other disciplines. Some prior applications of the 5E model include: curriculum materials such as unit plans and lesson plans; comprehensive educational products such as curriculum frameworks and assessment guidelines; teacher professional development programs; and informal education [5]. Specifically, when teachers implement the instructional material with a level of fidelity that is medium to high, the 5E model has been shown to produce a direct effect on student learning achievement [21]. When the 5E model is subject to an unsystematic implementation, lacking deliberate structure and purposeful organization, it may not yield the intended results [21]. Since the 5E instructional model engenders active participation in inquiry, it can produce durable student learning [7]. Teachers believe that as students perform laboratory activities, experiments, and investigations, their understanding of the learned concepts evolves from 'discrete to concrete' [7]. Moreover, with a high-fidelity implementation, the 5E model can enable students to enhance their questioning, observation, and decision-making skills [8]. An application of the 5E model in instruction can also lead to higher-order thinking [22] and improved student attitudes toward science [9, 10]. The integration of interactive and hands-on activities in the 5E instructional model promotes inquiry-based learning that leads to better transfer of knowledge and higher engagement [8]. The 5E model is deemed practical [7] when specific disciplinary content, instructional methodology, and classroom setting, among others, can suitably profit from this model, which can additionally guide in the development of a constructivist lesson that can effectively remedy student misconceptions [23].

3. Literature review on integrating the 5E model in robotics-based learning

A judicious integration of appropriate technology in teaching and learning promises to enhance student skills in scientific inquiry and problem-solving [11, 12]. Thus, it is critical to embed opportunities for teachers to enhance their STEM pedagogy and concurrently develop their technological, pedagogical, and content knowledge (TPACK) [24]. Equipped with the 5E framework, teachers can gain a better understanding of complex interactions between technology, pedagogy, and content and thus improve their knowledge and skills for designing technology-enhanced lessons [12]. This framework can be an effective approach to design lessons and deliver instructions for robotics-based activities that can enhance interest and motivation among students toward learning. Following is a brief description of several studies that have utilized the 5E model to design robotics-led classroom learning experiences.

A recent study by Guven et al. [25] utilized the 5E model to conduct Arduino-assisted robotics coding activities in a sixth-grade STEM summer elective course with 11 students (six girls and five boys). The study sought to find the effects of 5E-aligned robotics activities on the students concerning their science creativity, robotics attitudes, science motivation, ability to connect STEM learning with daily life, and attitudes toward science. The study posited that robotics coding activities can equip learners to develop perceptions for understanding the world around them and investigating solutions to real-life problems. The students received training in robotics coding and performed robotics activities incorporated in the 5E model for 80 minutes per week for nine weeks. Over two additional weeks, for pre-/post-course assessment, data collection was performed using several validated qualitative instruments and measures. During the elective course, supported by their teacher, the students performed various science experiments on topics such as electricity, environment, light, and sound, all having real-life relevance. Proceeding through the five phases of the 5E model, the students: engaged in brainstorming in response to teacher questions, for example, on energy efficiency, triggering their prior knowledge; explored robotics components, for example, solar panel, sensors, motors, etc., and algorithms using the Arduino platform to assemble and test robot operation; explained their understanding about various energy-related topics posed by the teacher; elaborated their learning through robotics coding projects for novel, self-selected, energy-efficient systems; and evaluated their robotics-based learning through reflections that were assessed using a rubric. The results of the study showcased positive outcomes wherein the students had improved creativity, attitudes toward robotics, and motivation toward science learning. Moreover, the study asserts that their 5E-aligned robotics coding activities can help learners form a better understanding of abstract concepts and apply them to their daily lives.

A quasi-experimental study by Hsiao et al. [26] recruited 70 sixth-grade students for a robotics-based learning activity that was spread over an 18-week duration, with two 40-minute sessions each week (including the pretest in the first week and the post-test in the last week). Based on the concept of ecosystem science and situated in the experience of the estuary area where the school was located, the students in both the control and experimental groups were assigned to design a "crab robot" that was tasked to "cross the road." All students participated in three stages of robot development that included exposure to block-based coding, knowledge about electronic components and sensors, and collaborative building of the robot. While the students in the control group learned about the robotics activities via traditional lectures, the students in the experimental group learned through the 6E model [27] (a revised 5E model with engage, explore, explain, *engineer*, *enrich*, and evaluate phases). Through the six stages, the experimental group students: engaged in learning via videos that sought to elicit their preconceptions; explored a worksheet-based game that built their understanding and knowledge structure; explained the learning activity by gaining an understanding of the components, concepts, and operation of the robot and coding; engineered solutions by applying their design and coding skills; enriched knowledge by applying their learning to solve problems; and evaluated their designs for iterative improvement to achieve the assigned task. For the students from the experimental group, the study results showed positive outcomes for computational thinking abilities, hands-on skills, learning motivation, and learning performance. Finally, the 6E-aligned robotics coding reinforced students' connections with real-life experiences and helped in building their interdisciplinary knowledge.

Yet another pilot study [28] that aimed to foster curiosity recruited 98 secondary and high school students from four schools to teach robotics using the 5E model. The students drew on their prior learning in answering "What is a robot?" during the engage phase, and then explored the physical components of the robot through assembly and programming. Next, they explained their understanding of the robotics concepts and components, then elaborated their learning by applying it to newer situations, and finally constructed and presented robots during the evaluation stage. This pilot study asserts that their approach can generate enthusiasm among students in using robots and can allow for autonomous exploration of concepts.

The above three studies showcase the positive influence of applying the 5E framework to robotics-based lessons for enhancing student engagement and learning effectiveness. Specifically, by embedding hands-on experiences, the 5E instructional model cultivated authentic contexts that enabled the students to link learning with their lived experiences and gain an interdisciplinary perspective of abstract concepts. Apart from positively influencing learning, the above studies demonstrated favorable outcomes in the form of improved creativity, enhanced attitude toward learning, and heightened enthusiasm for independent exploration of concepts.

4. Exemplar robotics-based science unit plans aligned with the 5E model and Next Generation Science Standards

Educational robotics aids in a deeper understanding of concepts, better retention of knowledge, and innovative thinking [29]. Specifically, for STEM concepts, robotics can act as a manipulative that can make the abstract ideas more accessible to all learners [13]. Incorporation of robotics in science learning offers opportunities for learners to interact with engineering artifacts, develop their ability and know-how to build robots, and apply science concepts to authentic situations arising from robotics. Such novel learning pathways can induce a conceptual change in students [30]. Moreover, robots are known to engender engagement in learners that can sustain and deepen their interest in STEM disciplines [30]. Thus, robotics-aided learning is aligned with the vision of the *Next Generation Science Standards* (NGSS) [31], specifically engineering practices [14], and can support the curriculum in reaching its intended outcomes [32].

Among a variety of factors impacting the classroom integration of any instructional model, one factor is the alignment of resulting plan and activities with the national standards in a purposeful and effective manner [30]. For example, to promote an effective understanding and exploration of science topics, the NGSS encourage teachers to adopt a three-dimensional model consisting of Science and Engineering Practices (SEP), Disciplinary Core Ideas (DCI), and Crosscutting Concepts (CCC) [30]. Below we highlight two examples of robotics-based 5E science lessons where the five phases align with the three dimensions of NGSS [31] and the Common Core State Standards for Mathematics [33] and Literacy. See Table A.6 in Appendix A for a lesson planning template that employs the 5E instructional model.

4.a. Scale of the solar system [32, 34]

The lesson in Table 9.2 utilizes a robotics-based scaled model to simulate the behavior of our solar system. The goal of the lesson is to record and analyze the orbital period data and arrange the order of planets based on their proximity to the Sun.

This robotics-enhanced learning activity seeks to embed an opportunity for the students to explore the scientific discovery process and develop their appreciation toward it. To do so, it makes use of a scaled model of the solar system that is studied through robotics explorations in a series of collaborative group activities involving investigations, data analysis, and knowledge synthesis. Under the 5E instructional model, the unit begins by activating students' prior knowledge of the solar system, for

TABLE 9.2 A unit plan on the solar system for middle school

Objectives: By the end of this unit, students will be able to *develop an appreciation toward the scientific discovery process* by

- simulating the behavior of the solar system using a small-scale model for it and
- developing skills in recording and interpreting observations.

Key questions:

- How can we learn about large systems even though we cannot observe them fully?

Assessment evidence: The students will

- *identify* the planets based on orbital periods through investigation and data analysis using robots;
- *apply* their learning from observations and data analysis to a new thinking problem; and
- *demonstrate* their learning through experimentation on a related problem of their choice.

Material required: A pre-programmed LEGO robot, a stopwatch, and a graphic organizer for each team of three to five students

Activity setup: The students produce a simulation with the LEGO robot to observe and record time taken by each individual planet, from the four selected planets, to complete one orbit around the Sun. They are then given the actual orbital period of Earth and asked to calculate the orbital periods of the other three planets based on the data collected.

NGSS Performance Expectations	**MS-ESS1-3 [31]** Analyze and interpret data to determine scale properties of objects in the solar system.	
Science and Engineering Practices (SEP)	**Disciplinary Core Ideas (DCI)**	**Crosscutting Concepts (CCC)**
Analyzing and Interpreting Data: Analyze and interpret data to determine similarities and differences in findings.	Earth and the Solar System: The solar system consists of the Sun and a collection of objects, including planets, their moons, and asteroids, that are held in orbit around the Sun under its gravitational pull on them.	Scale, Proportion, and Quantity: Time, space, and energy phenomena can be observed at various scales using models to study systems that are too large or too small.

Common Core	English Language Arts/Literacy—RST.6–8.1, RST.6–8.7 Mathematics—MP.2, 6.RP.A.1, 7.RP.A.2
	Unit Plan
Engage	*Opening activity—How are the planets arranged in the solar system?* • Discuss in groups: What do you know about the solar system? How are the planets arranged in the solar system in terms of their distance from the Sun? Which planet is closest to the Earth? Which planet is farthest from the Sun? • Provide the students with printed images of the planets and instruct them to place these around the Sun represented on a chart paper. • Discuss in groups: Remember a time when you gazed at the sky at night and in daytime, what key differences did you observe? Can you spot stars and planets in the sky without a telescope? If you were an astronomer in the past, how would you distinguish between stars and planets in the night sky? If today you are tasked to identify

(*Continued*)

TABLE 9.2 (Continued)

	planets in the night sky, how will you do so? Discuss your ideas with your group. How have your friends approached this challenge?
Explore	*Prompt the students to discuss in small groups* • What are some very small or large scale systems that are hard to observe? • What are some ways through which you can learn about them? Encourage the students to work in groups for generating their own questions about scaled models or arrangement of planets in the solar system that they want to examine as a part of this unit. *Robotics-aided activity—Record time taken by a planet to orbit around the Sun* This activity aims to create a scaled model of the solar system using a robot to make such observations accessible. • Position the robot on the chart paper and run the program for a planet of your choice. • Observe and record the time it takes for the robot to complete one full orbit. Record the observations using Table 9.2.a. • Run the LEGO robot programs for other planets of your choice and record the time taken for their full orbits. • Analyze the data and arrange the order of the planets based on their proximity to the Sun.

TABLE 9.2.A Record observations for the 'explore' activity of the learning unit plan

Planet	Observed time to orbit (secs)
Planet 1	
Planet 2	
Planet 3	
Planet 4	

Explain	Using videos and PowerPoint presentation • *Introduce the terms:* Planets, solar system, scaled systems, orbit, and orbital period. • *Discuss:* Student findings from the 'explore' activity. • *Explain:* Orbital period of the planet depends on its closeness to the Sun. The closer a planet is to the Sun; the shorter its orbital period.
Elaborate	*Group activity—Find orbital periods of planets* • Record in Table 9.2b the time taken by planets to orbit around the Sun in 'Earth days.' Note that the time required for Earth to orbit the Sun is 365 days.

TABLE 9.2.B Approximate observations to calculate orbital periods of planets

Planet	Observed time to orbit (secs)	Actual orbital period (Earth days)
Planet 1		365.2
Planet 2		
Planet 3		
Planet 4		

- In Earth years, calculate the time taken by each planet to orbit around the Sun. Record your results in Table 9.2.c.
- Compare and discuss responses with other groups.

TABLE 9.2.C Approximate observations to calculate orbital periods of planets in Earth years

Planet	Actual orbital period (Earth days)	Actual orbital period (Earth years)
Planet 1	365.2	1
Planet 2		
Planet 3		
Planet 4		

Group activity—Finding solution to the student-driven problem

- Based on their newly acquired learning, the students are provided time to conduct research on a related question that they generated during the 'explore' stage.
- The students work in groups to propose and devise possible solution strategies, conduct experiments, and obtain solutions.

Evaluate

- Provide actual orbital data of all the planets without disclosing the names of the corresponding planets. Using the data collected and analyzed from the activities above, ask the students to identify the planets in Table 9.2.d.
- *Thinking Problems*: Is one year on Mars longer or shorter than that on Earth? Calculate it in terms of Earth days. What is your age in terms of Mars years? What is it in terms of Earth years? What are the benefits and limitations of simulating a real-world system with a robotics-based scaled model?
- *Presentation*: Ask the students to showcase their investigations based on the question they generated in the 'explore' stage and their proposed experimental strategy to address it from the 'elaborate' stage. Ask them to explain how their question aligns with the theme of the unit and have them discuss the outcomes of their investigation.
- Ask the students to note their reflections in a learning journal.

TABLE 9.2.D Identify the planets

Planet	Actual Orbital Period (Earth days)	Name of the planet
A	3.652×10^2	
B	5.98×10^4	
C	8.8×10^1	
D	9.056×10^4	
E	4.331×10^3	
F	2.247×10^2	
G	1.0747×10^4	
H	6.87×10^2	
I	3.0589×10^4	

example, by engaging them in a discussion about stimulating questions concerning differences between planets and stars. To help reveal the current knowledge and understanding of the students, this phase also includes an activity about the order of planets in our solar system. The learning activities hook student attention by asking them to arrange the planets on a chart paper followed by a discussion (engage). Next, the students perform an activity with a LEGO robot wherein the robot simulates the behavior of a planet by orbiting around a selected point that represents the Sun. For the simulated scaled system, that is, for various planets, the students collect orbital periods using a timer and then they perform data analysis to generate findings (explore). In the spirt of the exploratory nature of inquiry, the students can be encouraged to generate a question or situation, aligned with the unit, and conduct investigations on it. This activity is followed by teachers' explanation of the new ideas and terminologies on the solar system, planetary revolutions, orbital periods, etc., that helps reinforce students' ideas and clears any misconceptions (explain). With the new understanding, the learners cooperatively [18] conduct further investigations and convert the small-scaled simulation data to the actual orbital period data (elaborate). In addition to working on the given scaled-model simulation, the students can use their learnings from this activity to perform research and experimentation on the questions they generate during the explore phase, facilitating transfer of knowledge and conceptual change. Finally, the learners respond to activities and thinking problems to assess their learning (evaluate). Moreover, they can be afforded opportunities to showcase experimental results for their self-generated question. As evidenced above, in line with the constructivist approach, throughout the unit the students must be encouraged to ask questions and construct ideas based on their discussions. Formative assessments and checks for understanding elements were embedded throughout the unit to enable the teacher in amending upcoming lessons for improving student learning.

In this unit, robotics as a pedagogical element affords the opportunity to make the learning tangible and visible. Specifically, the hands-on use of robotics allows learners to actively participate in exploring an abstract topic, which is otherwise unattainable due to the large scale of the solar system. For example, advanced observation and measurement technologies may not be accessible to learners to obtain the orbital periods of various planets. The robotics-based exploration renders opportunities to students to investigate the concept at their own pace and generate enthusiasm about an idea. Finally, the 5E instructional model aids learners in generating their own conceptions and elaborating their learning. For instance, this unit can also involve students to program their robots and work on other analogous problems that can be simulated via robotics.

4.b. Newton's first and second laws of motion [32, 35]

This unit (see Table 9.3) introduces to learners the concepts of balanced and unbalanced forces as well as Newton's first and second laws of motion using LEGO robot-based activities embedded in the 5E instructional model.

TABLE 9.3 A learning unit on Newton's first and second laws of motion

Objectives: By the end of this unit, students will be able to

- *identify* different forces working on a selected object;
- *describe* the effect of various forces on the motion of an object; and
- *explain* Newton's first and second laws of motion.

Key questions:

- What are Newton's first and second laws of motion?

Assessment evidence: The students will

- *create* a design for their own robot for an imaginary tug-of-war competition and
- participate in a thinking problem to *demonstrate their understanding* of Newton's first and second laws of motion.

Material required: Pre-programmed LEGO robots, balls, blocks, and small weights for each team of three to five students

Activity setup: Two sets of LEGO robots that will be modified during the activities for their construction, power level, and brick position.

NGSS Performance Expectations	**MS-PS2-2 [31]** Plan an investigation to provide evidence that the change in an object's motion depends on the sum of the forces on the object and the mass of the object.	
Science and Engineering Practices (SEP)	**Disciplinary Core Ideas (DCI)**	**Crosscutting Concepts (CCC)**
Planning and Carrying Out Investigations: Plan an investigation individually and collaboratively, and in the design: identify independent and dependent variables and controls, what tools are needed to do the gathering, how measurements will be recorded, and how many data are needed to support a claim.	Forces and Motion: The motion of an object is determined by the sum of the forces acting on it; if the total force on the object is not zero, its motion will change. The greater the mass of the object, the greater the force needed to achieve the same change in motion. For any given object, a larger force causes a larger change in motion.	Stability and Change: Explanations of stability and change in natural or designed systems can be constructed by examining the changes over time and forces at different scales.
Common Core	English Language Arts /Literacy—RST.6–8.3, WHST.6–8.7 Mathematics—6.EE.A.2, 7.EE.B.3, 7.EE.B.4	
	Unit Plan	
Engage	Show various images of objects experiencing forces and ask the students if forces on each object are balanced or unbalanced. Ask them to discuss their responses in groups.	
Explore	*Small group activity—Formulating a problem based on unbalanced forces* • Ask the students to work in groups to list real-life situations where they have witnessed unbalanced forces in action. • Encourage them to generate questions based on the phenomena and probe their peers and teacher for answers.	

(*Continued*)

TABLE 9.3 (Continued)

Activity—Robot Tug-of-War

- Engage the students in robot tug-of-war activities (see Figure 9.2) with three different robot specifications.

Activity 1: Two robots labeled A and B have both similar specifications of power level and construction.

Activity 2: Modify robot A by changing its power level to 80 and without altering its construction.

Activity 3: Modify robot B such that it has the same power level as robot A but its programmable brick position is altered or its weight is increased.

- Collect the results of the tug-of-war activities for three activities in Table 9.3.a.
- Discuss which robot won and why.

FIGURE 9.2 Robot tug-of-war [35].

TABLE 9.3.A Robot tug-of-war

	Robot A			**Robot B**		
	Initial position	Final position	Distance travelled	Initial position	Final position	Distance travelled
Activity 1						
Activity 2						
Activity 3						
	Average distance travelled			Average distance travelled		

Explain Using robots as examples, discuss the following concepts and illustrate them with free-body diagrams to support explanations.

- Different types of forces: gravitational, normal, frictional, pulling, and pushing
- Newton's first law of motion
- Balanced and unbalanced forces
- Effect of different types of wheels on the robot
- Effect of adding mass on the object

Elaborate *Activity 1– Inertia—Newton's first law of motion*

- As shown in Figure 9.3, put one or more blocks on the robot and run the robot suddenly such that a jerk is produced. Observe and record the outcome. Reason why this outcome happened. Discuss in groups.

FIGURE 9.3 Robot with weights [35].

Activity 2—Newton's second law of motion

- Move the robot without adding any load to it. Record its speed and distance traveled after 15 seconds.
- Place a ball or an object in front of the robot (see Figure 9.4). Do the same as above and observe and record the outcome.
- Gradually increase the weight of the ball. Now perform the above procedure and observe and record the change in robot's speed and distance traveled after 15 seconds.
- Observe and record the outcome. Reason why this outcome happened. Discuss in groups.

FIGURE 9.4 Robot carrying a ball [35].

(Continued)

TABLE 9.3 (Continued)

	Activity 3—Finding solution to the student-driven problem • Based on their learning about Newton's laws of motion from the robotics activity, the students are provided time to conduct research on a related question that they generated during the 'explore' stage. • The students work in groups to propose and devise strategies, conduct experiments, and obtain solutions. • The students observe and record similarities and differences between the robotics-based tug-of-war activity and their own investigations.
Evaluate	• *Design project:* Design your own robot to win a tug-of-war competition. What specific considerations would be a part of your design? How will you plan for more net force to win the tug of war? How does your learning from this activity translate to the questions you generated on a real-life situation? • Ask the students to record their reflections.

The unit on Newton's laws of motion begins by hooking student attention to the topic of balanced and unbalanced forces as well as helping teachers ascertain students' current understanding (engage). Then, the students participate in activities in which they conduct experiments on two robots and formulate interpretations about robot behaviors in different situations. The activities allow the learners to discover the notion of various forces through a tug-of-war between the two robots with similar and different specifications at different instances (explore). At this stage, the students can be given opportunities to identify real-life situations that occur due to unbalanced forces and ask questions about them. After preliminary investigations, teachers introduce the learners to new terminologies and concepts of Newton's laws that support the learners in validating their interpretations deduced from the former activity (explain). With new knowledge, the learners make connections and conduct additional investigations to study the effects of inertia and mass on robot behavior, thereby observing Newton's laws of motion in action (elaborate). They can be prompted to use their newly acquired knowledge to conduct investigations on the question of their choice identified in the 'explore' phase. This activity solidifies students' understanding as they begin to envision real-life application of these concepts. To demonstrate their learning, the students design their own robots to make efficient players for a tug-of-war competition (evaluate). Like the previous unit, formative assessments, classroom discussions on the ease of use of robots, and technology troubleshooting were planned ahead of time and embedded in the unit to ensure its purposefulness.

This robotics-enhanced 5E learning experience takes learners through various stages of constructing their learning about the topic and promotes spatial reasoning and encourages critical thinking. Specifically, the use of a robot as a learning tool allows learners to examine how manipulation of multiple variables in the system can affect the center of gravity. Students utilize their prior knowledge of the centers of mass and gravity to manipulate the programmable brick of the robot to various locations and hypothesize what they think will happen when pulling another robot in the tug of war. Such interrelated knowledge can make the learning more exciting and engage the students in a process that can help them relate what they are learning in class to real-life scenarios.

Finally, see Appendix A, Table A.7 for a learning unit on genetic mutation and Table A.8 for a lesson plan on speed, distance, and time, both of which use robotics explorations with the NGSS-plus-5E framework. Several additional NGSS-plus-5E robotics lessons are listed in Table B.2 in Appendix B.

5. Implementing the 5E instructional model

The widespread use and popularity of the 5E model has resulted in the identification of factors that can assist in making the model more robust and effective in its implementation. In fact, the original proponent of the 5E model, Rodger W. Bybee, collected his reflections in an article [19] whose recommendations are summarized in the following subsection. For example, Bybee [19] recommends that a 5E-based instructional unit is most effective when its implementation is spread over two to three weeks, instead of a single week, to provide sufficient opportunities to facilitate conceptual change in learners. The role of teacher professional development and integration of the model with NGSS also impacts the implementation of the 5E model and a summary of each of these factors is also included in the remaining two subsections.

5.a. Recommendations for implementation [19]

Even though the 5E instructional model is presented and explained as if evolving along a linear trajectory, it is actually a cyclic model that is flexible and dynamic. It is recommended that the model phases be reiterated as necessary to impart a deep learning experience to students. For example, students may cycle through multiple explore and explain experiences prior to being transitioned to the elaboration phase. The robotics-based science units above were flexible and the teachers revisited the data collection and analysis stages multiple times before the students could make conclusions that were robust and met the lesson objectives. The model advocates that students be first introduced to a new concept since this approach offers an opportunity for a complete learning cycle.

The 5E instructional model is most effective when implemented in a unit of two to three weeks duration wherein each phase of the model consists of a distinct lesson. The effectiveness of the model is limited if all five phases are used in a single lesson since it curtails the time and opportunities for a conceptual change to occur. For example, the robotics-based science units in Section 4 were planned for a duration of three to five distinct lessons. An attempt to cover all the five phases in one standalone lesson may limit students in actually engaging in an inquiry approach and restrict their discovery process. Moreover, if excessive time is devoted to one or more phases, then students may forget what they learned in prior phases and the overall structure may become ineffective.

To have the greatest impact on learning, no phase must be omitted and their order should not be altered. When some instructional models start directly with the explain phase, omitting the engage or explore phases, or some others omit the elaborate phase, they may fail to achieve desired student learning outcomes. The design of the 5E instructional model is informed by how students learn, and the appropriate use of the phases promotes the transfer and application of knowledge. The robotics-based science units are consistent with this model and render an opportunity for students to deeply engage with the content.

The engage phase should not be confused with a pre-assessment or a diagnostic test. Instead, this phase is an opportunity for teachers to draw out students' current conceptions of knowledge in an informal way. When applicable, activities in a 5E learning sequence should be aligned with the three dimensions of the NGSS, namely SEP, DCI, and CCC. The explain phase of the model can be used to introduce vocabulary and scientific skills in an understandable and clear way. For instance, the units of Section 4 explain the concepts of orbital periods and laws of motions after preliminary activities.

Formative assessment represents a continuous improvement process and is encouraged throughout the instructional and learning plan to make iterative refinements to the plan as per student needs. The evaluate phase highlights the summative assessment and should be conducted toward the end of the unit. Although not presented in the units above, formative assessments in the form of short quizzes, small group discussions, data analysis reviews, essays, and oral check for understanding were embedded as a part of each of the lessons so that the teachers can make necessary amendments to improve student learning. For example, under teacher observation, small group discussions in the explore phase of the solar system lesson (see Table 9.2) can help in addressing misconceptions and reinforcing the topics, correct definitions, etc. Similarly, reviewing data collection and analysis worksheets completed by students in the elaborate phase of the two lessons in Section 4 can inform teachers about students' ability to apply content knowledge from the robotics activity.

Embedding multiple opportunities for generative discussions and encouraging inquiry through questioning in the units can allow students to demonstrate their learning. With the use of discussion questions, prepared in advance, students can be prompted to provide evidence and arguments in support of their claims as well as applying their knowledge to a new situation, all of which can support the integration of the three dimensions of NGSS [30]. Moreover, student responses to interactive classroom discussions can serve as a formative assessment by illustrating their understanding about a topic and eliciting their misconceptions before the teacher transitions to the next phase of the lesson. Finally, frequent discussion on the use of robots in the classroom and technology troubleshooting are important elements of robotics-enhanced units since they assists in surfacing any logistical limitations that students may be facing in achieving the intended learning goal.

5.b. Designing teacher professional development for robotics-based lessons integrating the 5E model

Although inquiry-based teaching is widespread in education communities, teachers find it challenging to design such units [4]. A structured teacher professional development model can assist educators in mastering challenging teaching strategies and in changing instructional behavior, especially pertaining to topics in science and engineering practices [14]. Additionally, since instructional materials with high fidelity to the 5E model can produce higher student learning gains [21], collaborative teacher professional development can serve as one mechanism to achieve such an outcome. It has the potential to enhance students' scientific inquiry and problem-solving skills since the knowledge and skill to effectively incorporate the 5E model can help improve teachers' TPACK [12]. The 5E framework can be utilized as the instructional methodology of professional development programs and intentionally modeled during professional development to permit sufficient time for teachers to process each of the phases.

Specifically, a robotics-based professional development program can be designed using the 5E instructional model such that it [32]: engages teachers by integrating examples of successful classroom implementation of standards-aligned 5E lessons to awaken teacher interest and enthusiasm in the model [32] (engage); allows teachers to gain an initial understanding of the elements of the 5E model and encourages them to explore sample 5E-based lesson plans and peruse research related to the 5E model (explore); introduces teachers to various phases of the 5E model and explains its key implementation features (explain); supports teachers as they seek to embed science standards in their lessons (explain); assists teachers to acquire knowledge and skills concerning relevant technology tools (explain); allows teachers to gain hands-on experiences by encouraging them to develop and model their own 5E-based lessons (elaborate) [32]; encourages teachers to compare and contrast the 5E instructional model with other comparable models (elaborate); and provides teachers opportunities to showcase their newly acquired knowledge and obtain feedback from facilitators and peers (elaborate and evaluate). As an example, after the exposure to the hands-on application of the 5E model in lesson creation during the professional development program, a teacher stated [32] "My frame of mind for lesson planning is more structured, as I think of each of the 5E components I need to address."

In line with the collaborative aspect of the 5E model, professional development facilitators can also promote peer learning and give opportunities to teachers to interact with each other; build a network of teachers to share reflections, best practices, and concerns; and provide future avenues for teachers to become mentors to new participants [32]. The impact of discussions and feedback on the design of lessons was specifically noted in the professional development study of [32] where a participating teacher acknowledged that "a great deal of discussion . . . helped us develop an even better lesson than we [had] initially written." To best incorporate the standards in the learning design, teachers must be given professional development opportunities to deeply explore the standards alongside the instructional model. Such a professional development can help teachers to understand and articulate the objectives of the standards and align them with the pedagogical elements [30]. For instance, active participation in curriculum development integrating the three dimensions of NGSS [32] allowed a teacher to "have a much better understanding [of the format] now."

6. Conclusion

Using the 5E instructional framework can help the lessons shift away from memorization and consumption of facts to creation of hands-on and collaborative [18] learning experiences. The five phases of the model–engage, explore, explain, elaborate, and evaluate–are built on prior research, display coherence of instructional flow of the lesson, and permit ease of implementation for educators. The sustained use of the model can help in better retention of concepts, improved perception toward science, and higher student achievement. To render high fidelity of implementation for the model, educators ought to be provided opportunities for professional development and collaborative learning. Additionally, robotics-aided lessons can be effective in generating scientific inquiry among students and promoting fundamental ideas of science. Thus, they must be integrated appropriately in lessons. Overall, incorporating the 5E model along with robotics can be an effective and meaningful way to design constructivist science lessons and can lead to a strong scientific foundation for learners.

7. Key takeaways

- Inquiry-based learning technique allows students to formulate and discover new concepts by asking questions; conducting observations; investigating phenomena; performing comparisons; making predictions; using tools to collect, analyze, organize, and interpret data; drawing conclusions and framing explanations; and communicating results [16, 36]. Inquiry-based science curriculum has been shown to engender active engagement, remove focus from rote memorization, and promote learning for understanding effectively [3].
- The 5E framework is an inquiry-based instructional model consisting of five phases, namely, engage, explore, explain, elaborate, and evaluate.
- The 5E model can lead to better scientific reasoning abilities and improved attitudes toward science [5, 9, 10, 22].

- For STEM concepts, robotics can act as a manipulative that can make abstract concepts more concrete and accessible to all learners [13]. Robotics offers opportunities for learners to interact with engineering artifacts, develop their ability to build robots, and apply science concepts to authentic situations. Such novel pathways can lead to a conceptual change [30].
- To increase the effectiveness of implementing the 5E model [19], educators must allow the learning cycle to be flexible and dynamic, use the model in a unit of two to three weeks duration instead of a single lesson, not omit any phases, and not confuse the 'engage' phase with a pre-assessment or a diagnostic test.
- The teachers should be provided with ample professional development opportunities to become familiar and comfortable with the inquiry-based 5E framework.

References

1. Symonds, W.C., Schwartz, R., and Ferguson, R.F., *Pathways to Prosperity: Meeting the Challenge of Preparing Young Americans for the 21st Century*. 2011, Cambridge, MA: Harvard University Graduate School of Education. p. 1–53.
2. Brickhouse, N., Bringing in the outsiders: Reshaping the sciences of the future. *Journal of Curriculum Studies*, 1994.26(4): p. 401–416.
3. National Research Council, *Inquiry and the National Science Education Standards: A Guide for Teaching and Learning*. 2000, Washington, DC: National Academy Press.
4. Duran, L.B. and Duran, E., The 5E instructional model: A learning cycle approach for inquiry-based science teaching. *Science Education Review*, 2004.3(2): p. 49–58.
5. Bybee, R.W., Taylor, J.A., Gardner, A., Scotter, P.V., Powell, J.C., Westbrook, A., and Landes, N., *The BSCS 5E Instructional Model: Origins and Effectiveness*. 2016, Colorado Springs, CO: Office of Science Education, National Institutes of Health.
6. Piaget, J., Piaget's theory, in *Carmichael's Manual of Child Psychology*. 1970, New York, NY: Wiley.
7. Metin, M., Coskun, K., Birisci, S., and Yilmaz, G.K., Opinions of prospective teachers about utilizing the 5E instructional model. *Energy Education Science Technology Part B: Social and Educational Studies*, 2011.3(4): p. 411–422.
8. Dodge, M.M., *The Effect of the 5E Instructional Model on Student Engagement and Transfer of Knowledge in a 9th Grade Environmental Science Differentiated Classroom*, Master of Science—Science Education, Montana State University July 2017, (Thesis).
9. Akar, E., *Effectiveness of 5E Learning Cycle Model on Students' Understanding of Acid-base Concepts*, Masters in Science, Middle East Technical University The Graduate School of Natural and Applied Sciences 2005, (Thesis).
10. Tinnin, R.K., *The Effectiveness of a Long-term Professional Development Program on Teachers' Self-efficacy, Attitudes, Skills, and Knowledge Using a Thematic Learning Approach*, The University of Texas at Austin 2000, (Thesis).
11. McFarlane, A. and Sakellariou, S., The role of ICT in science education. *Cambridge Journal of Education*, 2002.32(2): p. 219–232.
12. Mustafa, M.E.I., The impact of experiencing 5E learning cycle on developing science teachers' technological pedagogical content knowledge (TPACK). *Universal Journal of Educational Research*, 2016.4(10): p. 2244–2267.

13. Brill, A.S., Elliot, C.H., Listman, J.B., Milne, C.E., and Kapila, V., Middle school teachers' evolution of TPACK understanding through professional development, in *Proceedings of ASEE Annual Conference and Exposition*. 2016, New Orleans, LA; Available from: https://peer.asee.org/25720.
14. You, H.S., Chacko, S.M., and Kapila, V., Teaching science with technology: Scientific and engineering practices of middle school science teachers engaged in a robot-integrated professional development program (Fundamental), in *Proceedings of ASEE Annual Conference and Exposition*. 2019, Tampa, FL; Available from: https://peer.asee.org/33353.
15. Rutherford, F.J. and Alhgren, A., *Project 2061: Science for all Americans*. 1990, New York, NY: Oxford University Press.
16. National Research Council, *National Science Education Standards*. 1996, Washington, DC: National Academies Press.
17. Atkin, J.M. and Karplus, R., Discovery or invention? *The Science Teacher*, 1962.29(5): p. 45–51.
18. Johnson, D.W. and Johnson, R.T., *Learning Together and Learning Alone*. 1987, Englewood Cliffs, NJ: Prentice Hall.
19. Bybee, R.W., The BSCS 5E instructional model: Personal reflections and contemporary implications. *Science and Children*, 2014.51(8): p. 10–13.
20. Tanner, K.D., Order matters: Using the 5E model to align teaching with how people learn. *CBE Life Sciences Education*, 2010.9(3): p. 159–164.
21. Taylor, J., Scotter, P., and Coulson, D., Bridging research on learning and student achievement: The role of instructional materials. *Science Educator*, 2007.16(2).
22. Boddy, N., Watson, K., and Aubusson, P., A trial of the five Es: A referent model for constructivist teaching and learning. *Research in Science Education*, 2003.33(1): p. 27–42.
23. Tural, G., Akdeniz, A.R., and Alev, N., Effect of 5E teaching model on student teachers' understanding of weightlessness. *Journal of Science Education and Technology*, 2010.19(5): p. 470–488.
24. Koehler, M.J., Mishra, P., and Cain, W., What is technological pedagogical content knowledge (TPACK)? *Journal of Education*, 2013.193(3): p. 13–19.
25. Guven, G., Kozcu Cakir, N., Sulun, Y., Cetin, G., and Guven, E., Arduino-assisted robotics coding applications integrated into the 5E learning model in science teaching. *Journal of Research on Technology in Education*, 2022.54(1): p. 108–126.
26. Hsiao, H.-S., Lin, Y.-W., Lin, K.-Y., Lin, C.-Y., Chen, J.-H., and Chen, J.-C., Using robot-based practices to develop an activity that incorporated the 6E model to improve elementary school students' learning performances. *Interactive Learning Environments*, 2022.30(1): p. 85–99.
27. Barry, N., The ITEEA 6E learning byDeSIGN™ Model. *Technology and Engineering Teacher*, 2014.73: p. 14–19.
28. Buselli, E., Cecchi, F., Dario, P., and Sebastiani, L., Teaching robotics through the inquiry based science education approach, in *Proceedings of International Workshop Teaching Robotics Teaching with Robotics: Integrating Robotics in School Curriculum*. 2012, Riva del Garda (TN), Italy. p. 192–193.
29. Deliberto, B.R., *Robotics and Inquiry: Addressing the Impact on Student Understanding of Physics Concepts (Force and Motion) from Select Rural Louisiana Elementary Students through Robotics Instruction Immersed within the 5E Learning Cycle Model*, Louisiana State University Doctoral Dissertations, 2014, (Thesis).

30. You, H.S., Chacko, S.M., Rajguru, S.B., and Kapila, V., Designing robotics-based science lessons aligned with the three dimensions of NGSS-plus-5E model: A content analysis (Fundamental), in *Proceedings of ASEE Annual Conference and Exposition*. 2019, Tampa, FL; Available from: https://peer.asee.org/32622.
31. NGSS, *Next Generation Science Standards (NGSS): For States, By States*. 2013, Washington, DC: The National Academies Press; Available from: www.nextgenscience.org/.
32. Ghosh, S., Krishnan, V.J., Rajguru, S.B., and Kapila, V., Middle school teacher professional development in creating a NGSS-plus-5E robotics curriculum (Fundamental), in *Proceedings of ASEE Annual Conference and Exposition*. 2019, Tampa, FL; Available from: https://peer.asee.org/33108.
33. Common Core State Standards Initiative, *Common Core Standards for Mathematics*. Common Core Standards Initiative. 2010; Available from: https://learning.ccsso.org/wp-content/uploads/2022/11/Math_Standards1.pdf.
34. *Scale of the Solar System: NGSS Lesson Plan*; Available from: http://mechatronics.engineering.nyu.edu/pdf/scale-of-solar-system.pdf.
35. Chacko, S.M., *Forces and Interactions*. 2018; Available from: http://mechatronics.engineering.nyu.edu/pdf/forces-and-interactions.pdf.
36. Uno, G., *Handbook on Teaching Undergraduate Science Courses: A Survival Training Manual*. 1999, Independence, KY: Thomson Custom Publishing.

APPENDIX A

Related information from 'Teaching STEM with Robotics' project

This appendix consists of a variety of related information that was developed as a part of the 'Teaching STEM with Robotics' education research project. While some included items encapsulate important aspects of project design, others constitute suggestions and recommendations, and still others offer readily adoptable templates. Education researchers and classroom teachers may both find these items to be of value in developing and enacting their own robotics-based educational curriculum and teaching practices. The items related to project timeline and activities, surveys, instructional practices and models, lesson templates, sample lessons, etc., have all been co-created through a partnership involving engineering and education researchers and middle school teachers. The sample lesson plans were developed primarily for Grades 6–8, are aligned with the three-dimensional (3D) learning framework of the Next Generation Science Standards (NGSS) [1], embed Common Core State Standards (CCSS) for Mathematics [2], employ project-based learning, and use LEGO Mindstorms as the robotics artifact. These lessons should be appropriately customized according to the intended grade-level use and student background. The following items are included in this appendix:

1. A summary timeline for the 'Teaching STEM with Robotics' project
2. A comparison of design-based research with other research methodologies
3. Sample items from the TPACK self-efficacy and TPACK awareness surveys
4. Instructional practices for successful robotics lessons
5. A lesson planning template using Bloom's taxonomy
6. Characteristics of the 5E instructional model
7. A 5E lesson planning template
8. A learning unit on genetic mutations
9. A lesson plan on speed, distance, and time

1. A summary timeline for the 'Teaching STEM with Robotics' project

The six-year-long project employing robotics to facilitate teaching and learning was co-created in partnership with teachers using the principles of design-based research (DBR). It was founded on the theoretical underpinnings of technological, pedagogical, and content knowledge (TPACK), features of effective professional development, 5E instructional framework, project-based learning, and cognitive apprenticeship among others. The project impacted learning of 44 teachers in 22 middle schools, catering to over 2,000 students. See Table A.1 for a summary of the project activities.

TABLE A.1 Summary of project activities

Timeline	*Fall and Spring (Sept–May)*	*Summer (June–August)*
Year 1 (2014–15)	Pilot development of robotics-based science and math lessons via informal discussions with 16 teachers from a K-12 science, technology, engineering, and mathematics (STEM) education program.	Collaborated with two pairs of science and math teachers to pilot a three-week-long professional development program to improve the lessons and professional development model.
Year 2 (2015–16)	Two pairs of teachers from the year 1 pilot professional development program implemented the curriculum in their classrooms.	Conducted a three-weeklong workshop with 20 teachers using DBR and consistent with situated learning and TPACK. Teachers suggested refinements to the lessons and workshop model.
Year 3 (2016–17)	Supported the 20 teachers from the year 2 summer workshop in implementing robotics-based lessons in their classrooms. Facilitated on-going analysis to address the project's research goals.	Conducted a three-weeklong workshop with 23 teachers using DBR and consistent with situated learning and TPACK. Teachers suggested refinements to lessons and workshop model.
Year 4 (2017–18)	Supported the 23 teachers from the year 3 summer workshop in implementing robotics-based lessons in their classrooms. Facilitated on-going analysis to address the project's research goals.	During a three-week summer workshop, under the social capital model, the project team co-developed NGSS-plus-5E aligned robotics-based lessons with six master teachers and conducted a workshop with 11 new teacher participants to test and revise the lessons.
Year 5 (2018–19)	Supported the six master teachers from the year 4 summer workshop to integrate new NGSS-plus-5E robotics-based lessons into their classrooms for testing and feedback.	During a three-week summer workshop, the project team co-developed NGSS-plus-5E aligned lessons with three master teachers and conducted a workshop with 20 new teacher participants to test and revise the lessons.

(*Continued*)

TABLE A.1 (Continued)

Timeline	*Fall and Spring (Sept–May)*	*Summer (June–August)*
Year 6 (2019–20)	Teachers continued to implement NGSS-plus-5E robotics-based lessons in their classrooms. Project personnel continued analysis of research findings and dissemination of outcomes through scholarly journals, conference proceedings, and project website.	

2. A comparison of design-based research with other research methodologies

Design-based research (DBR) appears similar to other research methods due to some common characteristics like the iterative refinement of research, collaboration with practitioners in authentic settings, etc. Yet, it is distinct from other approaches and some of those differences are listed below in Table A.2.

TABLE A.2 Comparisons between design-based research and other methodologies

Characteristics of other methodologies	*Characteristics of design-based research*
1. Laboratory research *versus* Design-based research [3]	
• "Contaminating effects" are averted by obviating the diversions and disruptions that occur in real classrooms.	• Complexities entailed in a real classroom are present since the research is conducted in local environments.
• Participants are immersed in learning tasks and are socially isolated.	• Participants are placed in real settings that involve natural distractions and they interact with each other.
• Focus is on a single dependent variable.	• Even as multiple dependent variables exist, relationships between each of them is neither amenable to examination nor the focus of study.
• Material is presented in a recipe-like manner and the research findings are well documented.	• The processes start with planned procedures and iterative refinements are performed as needed.
• One or a few hypotheses are tested thoroughly.	• Myriad aspects that define design in practice are considered.
• Researchers are the sole decision-makers.	• Participants are involved at different stages of ideation, implementation, analysis, and revision.
2. Randomized controlled studies *versus* Design-based research	
• May judge the effectiveness of an intervention too early [4].	• Is conducted in contextual settings over long periods of time to accurately check for the efficacy of an intervention [4].
• May fail to acknowledge contextually dependent factors [4].	• Is contextually embedded and accounts for locally relevant factors [4].
• Lacks the ability to reveal causality between instructional processes *versus* learning outcomes [4].	• Can produce feasible causal connections [4].

Characteristics of other methodologies	*Characteristics of design-based research*
• Can be difficult to form similar composition treatment and control participant groups [5, 6].	• No compositional constraints on who to include in participant group [6].
3. Formative evaluation *versus* Design-based research	
• Is an evaluation method [7]. • Its primary goal is to improve the practice of design [7, 8]. • Consistent connection between the design intervention and theories is missing and the focus is on evaluation instead [8].	• Is a research paradigm [7]. • The primary goal is to generate theories [7]. • The process of connecting design intervention with existing theory is consistent through the phases [8].
4. Action research *versus* Design-based research	
• Begins with practitioners initiating the research process and then researchers assisting in facilitating the procedures [7, 9]. • Its goal is to benefit both research and practice by addressing practitioner's urgent and specific concerns and thus it can be greatly contextual [10].	• Begins with researchers deciding the problem to be addressed and then practitioners getting involved in the design, implementation, and analysis [7, 9]. • Its goal is to generate theories that can address pragmatic and real-life challenges and benefit the larger education community [7, 9].

3. Sample items from the TPACK self-efficacy and TPACK awareness surveys

For the studies of [11, 12] discussed in Chapter 5, a self-efficacy survey was adapted from a validated instrument in [13] that included 75 items with a five-point Likert scale. Specifically, guided by [14], the self-efficacy instrument of [13] was reformulated to obtain the measures of teacher self-efficacy for the seven components of TPACK along the four factors of confidence, motivation, success, and anxiety to yield 28 dimensions (e.g., TK confidence, . . ., TCK motivation, . . ., TPK success, . . ., and TPACK anxiety). The various TPACK items were altered to align with the educational robotics setting. The resulting self-efficacy survey included 116 items across 28 dimensions to be rated on a scale of 0 (low) to 100 (high). See some sample items below and [11, 12] for additional details.

Sample items from the TPACK self-efficacy survey [11, 12]: For each of the seven components of TPACK, one sample item is provided below. For each item, ratings to the four factors of confidence, motivation, success, and anxiety are elicited from the respondents.

Technological Knowledge (TK): Learn new technologies such as LEGO robotics

Pedagogical Knowledge (PK): Adapt teaching to class/student needs

Content Knowledge (CK): Use science/math thinking in problem-solving

Technological Pedagogical Knowledge (TPK): Adapt technology such as LEGO robotics to enhance teaching and learning

Technological Content Knowledge (TCK): Know about technologies such as LEGO robotics that can be used for understanding and doing science/math

Pedagogical Content Knowledge (PCK): Teach effectively to promote student learning of science/math

Technological Pedagogical and Content Knowledge (TPACK): Integrate robotics, science, and mathematical ways of thinking in lesson planning

Next, for the study of [15] discussed in Chapter 6, TPACK awareness and TPACK self-efficacy surveys were administered to investigate teachers' understanding and perceived importance of the TPACK framework as well as their TPACK self-efficacy in using robotics as a teaching tool. To ascertain the level of teacher awareness for the technology, pedagogy, and content domains of the TPACK construct, a questionnaire with eight items was designed and administered [15]. The validated self-efficacy instrument of [13] was adapted in the study of [15]. Specifically, the self-efficacy survey of [15] included a total of 26 items in the seven TPACK domains and was self-rated by teachers on a seven-point Likert scale. Below are some sample items from these two surveys.

Sample questions from the TPACK awareness survey [15]:

What technological, pedagogical, and content knowledge do you ideally require to plan and effectively teach the lessons using robotics?

What is the relative importance of the technological, pedagogical, and content knowledge for planning and effectively teaching the lessons using robotics?

What are the factors that may affect the requirements and relative importance of the technological, pedagogical, and content knowledge for effectively teaching the lessons using robotics?

Sample items from the TPACK self-efficacy for robotics lessons survey [15]:

Technological Knowledge: Have the technical skills to teach robotics-based lessons.

Content Knowledge: Have sufficient knowledge about mathematics/science required for middle school grades.

Technological Pedagogical and Content Knowledge: Can determine the requirements of technological, pedagogical, and content knowledge for my robotics-focused lessons.

4. Instructional practices for successful robotics lessons

To successfully implement robotics in classrooms the following list of recommendations (see Table A.3) has been curated through learnings from actual classroom enactment of robotics-enhanced lessons. These recommendations constitute effective practices to help: plan for prerequisite learning, enact classroom instruction, and develop positive perceptions for using robotics as a learning tool in classrooms.

TABLE A.3 Recommendations for implementing successful robotics lessons

1	**Develop prerequisite knowledge [16]**
1.1	Provide training sessions to teach students engineering concepts of robotics such as actuation, sensing, and control, as well as to demonstrate the use of basic laboratory instruments. Trainings can also be an avenue to discuss general social, behavioral, and managerial attitudes.
1.2	Introduce students to programs, lessons, and activities that foster computational thinking skill development. Active, hands-on, and inquiry-based programs that allow learners to think critically, problem-solve, conduct investigations, analyze data, participate in block-based programming, etc., can enable them to participate in robotics activities.
1.3	Assess learners on their status of prerequisite knowledge when designing lesson objectives. Such an assessment is desirable since lack of prerequisites may lower the effectiveness of robotics-based lessons.
2	**Plan for the lesson [16]**
2.1	Ascertain relevant situated scenarios so that the instructional design uses robotics as a pedagogical tool as opposed to a technocentric tool.
2.2	Scrutinize the lessons, robot's activities, and situations to ensure they do not lead to misconceptions, cause safety concerns for students, or require long durations for implementation.
2.3	Provide sufficient teacher professional development to impart knowledge, skills, and confidence to independently implement a robotics-based lesson.
2.4	Plan for class logistics and management including time duration, troubleshooting support for robots, arranging required hardware and software, planning student groups, etc.
3	**Ascertain student misconceptions [17]**
3.1	Identify and address student misconceptions using robotics-based activities. Use inquiry-based learning, discussions, entry slips, or lesson frameworks such as 5E to surface misconceptions.
4	**Identify suitable topics for robotics activities [17]**
4.1	Utilize the following checklist: • The topic has a scope for performing hands-on activities that incorporate robotics and facilitate improved student learning. • The integration of robotics endows students with opportunities to connect their learning with real-world applications and understand its significance for their future education and careers. • Abstract concepts, especially in science, that are challenging to visualize with traditional instructional methods can be best catered by robotics-based integration. • The use of robots effectively contributes to the learning objective through various aspects of the lesson instead of serving solely as an engagement tool. • The integration of robots must be aligned with appropriate grade level and learning standards. • The lessons or topics should provide opportunities to students to apply their problem-solving skills.

(Continued)

TABLE A.3 (Continued)

5	**Engender opportunities for student engagement and motivation [17]**
5.1	Divide students into small groups and assign each group member different roles such as chief assembly engineer, chief designer, chief inventory manager, etc.
5.2	Introduce students to the purpose of using robots for the activities explicitly.
6	**Measure student progress through assessments [17]**
6.1	Utilize entry and exit tickets, pre- and post-tests, online assessments, observations, follow-up questions, self-assessment surveys and checklists, quality of hands-on activities, data collection, and analysis, and quality of student responses on assigned worksheets.
7	**Create classroom management structures [17]**
7.1	Organize instructional binders that contain all the relevant information regarding hardware, software, activities, data collection, assembly procedures, general troubleshooting ideas, etc., to reduce confusion and encourage independent problem-solving.
7.2	Provide worksheets to record data and perform calculations to help in streamlining learning and preventing mistakes.
7.3	Assign student roles such as those of a designer, programmer, timekeeper, note-taker, etc. Rotate the roles for each lesson.
7.4	Explicitly define classroom norms to keep a track of the material and maintain discipline.
8	**Positively influence student perceptions [18]**
8.1	Test the hardware and the software prior to the class and troubleshoot possible errors to avoid their occurrence in the classroom.
8.2	Plan such that robots are explicitly seen as an essential element to understand and visualize the concept at a deeper level.
8.3	Collaborate with engineers, researchers, or other educators to create lessons. These lessons can also be disseminated widely such that more teachers can benefit from them.

5. A lesson planning template using Bloom's taxonomy

Following is a lesson planning template (see Table A.4) created to assist teachers and teacher educators in developing a technology-enhanced lesson plan using the cognitive levels of Bloom's taxonomy.

TABLE A.4 A lesson planning template using the cognitive levels of Bloom's taxonomy

Objectives: *Students will be able to* (***action verb*** *from the cognitive domain of Bloom's taxonomy*) + (***noun*** *representing the knowledge and skills that students are expected to gain*).
Key questions:
Assessment evidence:
Material required:
Activities and lesson plan:

Opening

Knowledge
Recall of methods, processes, patterns, structures, or settings
Verbs: Write, list, label, name, state, define, . . .

Comprehension
Understanding what is being communicated and making use of this information
Verbs: Explain, summarize, paraphrase, describe, illustrate, . . .

Application
Applying the understanding of concepts to different situations
Verbs: Use, compute, solve, demonstrate, apply, construct, . . .

Analysis
Drawing relationships between elements to make the ideas clear
Verbs: Analyze, categorize, compare, contrast, separate, . . .

Synthesis
Building a structure and drawing meaning from various elements
Verbs: Create, design, hypothesize, develop, . . .

Evaluation
Making decisions, supporting views, and creating judgments about the material learned
Verbs: Judge, recommend, critique, justify, . . .

6. Characteristics of the 5E instructional model

Guided by [19], we outline in Table A.5 below the five phases of 5E instructional model in detail with respect to ideal behaviors of learners and corresponding instructional strategies. Moreover, we explicate learner and teacher actions that are inconsistent with particular phases and should be avoided in practice. For instance, the engage phase focuses on uncovering learners' prior knowledge conceptions and hooking them in learning. Hence, learners should engage in activities that showcase their curiosity and surface their current knowledge. Thus, in this phase, teachers must act as facilitators to elicit actions that reveal desired responses. In this phase, learners should not seek a single solution and teachers should refrain from answering questions or explaining the concept. Instead, teachers must direct their efforts to heighten curiosity of learners and support them in uncovering connections.

TABLE A.5 Characteristics of the 5E instructional model [19]

5E Phase	*Learner Behaviors Aligned with the Phase*	*Instructional Moves Aligned with the Phase*	*Actions Inconsistent with the Phase*
Engage *How will the teacher/curriculum task activate/ expose learners' prior knowledge?* *How will the teacher/ curriculum task hook learner attention?*	• Asks what, why, and how questions that exhibit curiosity in the topic and shares current understanding	• Ask questions to expose learners' prevailing conceptions and understanding • Support learners in discovering links to prior work • Engender interest and curiosity	• Learners work to obtain a single solution or ask teacher for the correct answer • Teacher uses didactic instruction, explains phenomena, gives definitions, and provides answers
Explore *How will learners take initiative and actively use materials to uncover information?*	• Tests estimates and theories and proposes new ones • Plans and conducts investigations • Applies diverse solution techniques • Compares and contrasts ideas with those of others	• Encourage and observe peer interactions • Ask questions to elicit learners' understanding and correct course if necessary • Provide clarifications	• Learners work aimlessly and stop upon reaching a solution • Teacher supplies information, solution strategy, answer, and thinking process
Explain *How will the learners share their findings?* *How will the teacher/curriculum task connect learner findings with the new definitions and concepts?*	• Shares knowledge and understanding with peers • Critically absorbs, questions, understands, and shares explanations • Compares current understanding with prior knowledge • Adjusts concepts and explanations with the emergence of new rationale	• Encourage learners to explain their conceptual understanding • Have learners defend their knowledge conception through evidence and clarification • Share concepts, models, definitions, and skills through varied means • Build on learners' current understanding to share new concepts	• Learners offer ill-conceived explanations and accept them, even when lacking connection to prior learning and devoid of any rationale • Teacher fails to seek learners' explanations, accepts explanations that lack justification, and introduces disconnected facts and ideas

5E Phase	*Learner Behaviors Aligned with the Phase*	*Instructional Moves Aligned with the Phase*	*Actions Inconsistent with the Phase*
Elaborate *How will learners apply their learning to a new situation?* *How will learner develop a deeper knowledge of the concept?* *How will learners apply higher order thinking skills?*	• Applies concepts, models, definitions, and skills in new but similar situations • Uses prior knowledge conception to formulate and pose questions, suggest solution techniques, make decisions, and design investigative methods • Makes rational conclusions from evidence • Draws conceptual connections between new and previous experiences	• Encourage learners to apply and extend previously learned concepts, models, definitions, and skills in new situations • Ask learners to provide additional evidence, explanation, or rationale • Reinforce learners' use of previously introduced terminology and descriptions • Ask questions that help learners draw rational conclusions from evidence and data	• Learners work aimlessly without purposeful application of previously learned concepts, evidence, or explanations • Teacher directs learners to answer through a solution routine
Evaluate *How will all learners demonstrate mastery of the lesson objective?* *How will learners have an opportunity to summarize the big concepts they learned?*	• Gives feedback to other learners • Assesses own growth and knowledge • Probes concepts deeply for additional learning or investigation • Answers open-ended questions using observations, evidence, and previously accepted explanations	• Ask open-ended what, why, how questions • Observe as learners demonstrate conceptual understanding • Use diverse assessments for evidence of learner learning • Embed self-assessment opportunities for learners	• Learners respond with memorized answers or make conclusions devoid of evidence or explanation • Teacher introduces new concepts or tests terminology, facts, and definitions

7. A 5E lesson planning template

Given in Table A.6 below is a lesson planning template designed to assist teachers and teacher educators in developing a technology-enhanced lesson plan using the 5E instructional model.

TABLE A.6 A lesson planning template using the 5E instructional model

Objectives: *Students will be able to* (***action verb*** *from the cognitive domain of Bloom's taxonomy*) + (***noun*** *representing the knowledge and skills that students are expected to gain*).
Key questions:
Assessment evidence:
Material required:
Activity setup:
NGSS Performance Expectations:

Science & Engineering Practices (SEPs)	Disciplinary Core Ideas (DCIs)	Crosscutting Concepts (CCCs)

Common Core:

Unit Plan	
Engage *How will the teacher/curriculum task activate/expose learners' prior knowledge? How will the teacher/curriculum task hook learner attention?* **Explore** *How will learners take initiative and actively use materials to uncover information?* **Explain** *How will learners share their findings? How will the teacher/curriculum task connect learner findings with the new definitions and concepts?* **Elaborate** *How will learners apply their learning to a new situation? How will learners develop a deeper knowledge of the concept? How will learners apply higher order thinking skills?* **Evaluate** *How will all learners demonstrate mastery of the lesson objective? How will learners have an opportunity to summarize the big concepts they learned?*	

8. A learning unit on genetic mutations

Below in Table A.7 we provide an eighth-grade science unit on genetic mutations [20] where students are assigned the role of genetic counselors. Using robots students examine genetic details of different patients and compare it with their health history to recognize any genetic disorders. This learning unit uses the 5E instructional model and is aligned with the 3D NGSS [1] and CCSS for English Language Arts and Mathematics.

TABLE A.7 Learning unit on genetic mutations

Objectives: Students will be able to simulate with a robot the method of recognizing genetic mutations in a person and compare the information from multiple sources to decide if the symptoms denote disorders with genetic or non-genetic basis.
Key questions: What is meant by a genetic mutation and what may be its effect?
Assessment evidence: Students will respond to a claim with their arguments, showcase evidence based on their investigations, and solve a specific problem.
Material required: A pre-programmed LEGO Mindstorms EV3 robot, videos on sickle cell anemia and cystic fibrosis, sample patient medical history cards, and sample disorder information cards.
Activity setup: Students will be provided LEGO robots with six programs corresponding to six patient conditions. To collect data, robots will be placed on a poster and a program will be executed to move it along chromosomes lined up in a row. Students will observe the robot as it reaches a specific location and indicates any error through an alarm and by displaying a number.
NGSS Performance expectations [1]:
MS-LS3–1: Develop and use a model to describe why structural changes to genes (mutations) located on chromosomes may affect proteins and cause harmful, beneficial, or neutral effects to the structure and function of the organism.

Science & Engineering Practices (SEPs)	Disciplinary Core Ideas (DCIs)	Crosscutting Concepts (CCCs)
Developing and Using Models Modeling in 6–8 builds on K–5 experiences and progresses to developing, using, and revising models to describe, test, and predict more abstract phenomena and design systems. Develop and use a model to describe phenomena.	**LS3.A: Inheritance of Traits** Genes are located in the chromosomes of cells, with each chromosome pair containing two variants of each of many distinct genes. Each distinct gene chiefly controls the production of specific proteins, which in turn affects the traits OF the individual. Changes (mutations) to genes can result in changes to proteins, which can affect the structures and functions of the organism and thereby change traits.	**Structure and Function** Complex and microscopic structures and systems can be visualized, modeled, and used to describe how their function depends on the shapes, composition, and relationships among its parts, therefore complex natural structures/systems can be analyzed to determine how they function.

(*Continued*)

TABLE A.7 (Continued)

Common Core:
English Language Arts/Literacy—RST.6–8.1, RST.6–8.4, RST.6–8.7, SL.8.5
Mathematics—6.SP.B.5, 7.RP.2, 7.RP.A.3

Unit Plan

Engage

Opening activity: Surfacing current understanding on chromosomal and genetic mutations

- Pair the students and show them depictions of human karyotypes. Students determine dissimilarities between karyotypes and discuss their understanding about having extra chromosomes.
- Provide students with a simple rationale and examples that merely the knowledge of karyotypes may not facilitate identification of all genetic disorders. Show videos of examples: sickle cell anemia and cystic fibrosis are disorders stemming from genetic mutation.

Explore

TABLE A.7.a Patient medical history cards

Patient 1	Patient 2	Patient 3
Isabella, a teenage girl, has difficulty breathing. She has encountered frequent coughs and unwarranted phlegm. Throughout her life, she has had to visit the hospital often due to lung infections.	Sebastian, a high schooler, often gets tired and fatigued. Yet, he wants to be on his school's basketball team. He has been asked to SUBMIT his most recent blood report. Previously, he was found to be resistant to malaria.	Stark's one-year-old baby recently stopped making sounds and is experiencing recurrent seizures. The baby is not responsive to people and cannot sit or crawl, something she was able to do previously.
Patient 4	**Patient 5**	**Patient 6**
Peter who is 30-year-old was let go from his job since he was frequently absent. He has since found it challenging to land another job. In addition to having a low appetite for food, he is experiencing poor quality of sleep. He has been withdrawing from his social circle and prefers staying home.	Victoria who is 40-year-old recently complained to her family physician about fatigue, muscle cramps, and lightheadedness. She is frequently experiencing heart palpitations as well.	Dalip who is 25-year-old is engaged to be married soon and recently began experiencing uncontrolled muscle spasms and inability to make decisions. His fiancé does not exhibit any of these symptoms. Both of them want to enquire about his family history.

Robotics-aided activity to compare data on mutations from multiple sources

- Provide students with posters having markings representing chromosomes and a set of medical history cards for six patients (see Table A.7.a).
- Ask students to place the robot on the start location marked on the poster and execute the robot code corresponding to a particular patient. A stop by the robot at any location along the genome of the patient signifies a mutation. Students record the chromosome number indicative of the mutation when the robot stops at an error (see Table A.7.b).
- Repeat the steps for each patient. Students note any predictions about patient diagnosis (see Table A.7.c).
- Provide students with six disorder information cards: Tay-Sachs disease, cystic fibrosis, depression, Huntington's disease, sickle cell anemia, and Down syndrome. Based on the information gathered from the robotics activity and the disorder information, ask students to note their medical diagnosis.
- Teachers can use the answer key shown in Table A.7.d to check student work.

TABLE A.7.b Table for students to gather data using robots

	Patient 1	Patient 2	Patient 3	Patient 4	Patient 5	Patient 6
Presence of mutation	☐ Yes ☐ No	☐ Yes ☐ No	☐ Yes ☐ No	☐ Yes ☐ No	☐ Yes ☐ No	☐ Yes ☐ No
Chromosome # with mutation						
Notes from brainstorming						

TABLE A.7.c Table for students to record their diagnosis

	Patient 1	Patient 2	Patient 3	Patient 4	Patient 5	Patient 6
Mutation on chromosome						
Possible genetic diagnosis						

TABLE A.7.d Answer key for the activity

Patient	1	2	3	4	5	6
Mutation on chromosome	7	11	15	n/a	n/a	4
Disorder	Cystic fibrosis	Sickle cell trait	Tay-Sachs disease	Depression	Dehydration	Huntington's Disease

(Continued)

TABLE A.7 (Continued)

Explain	*Discuss and provide new information* • Discuss diagnosis by different groups by asking students to present their evidence. • Explain about recessive disorders and that some of diseases they investigated fall under the category of recessive disorders. Such a disorder occurs in a person when they inherit two mutated genes, one from each parent. • Introduce the students to Punnett Square that indicates the likelihood of a child acquiring a recessive disorder due to their parents' genotypes. • Ask students to revise their responses from previous activity.
Elaborate	*Analyze data and group discussions* • Emphasize that mutations can yield anomalous proteins which can cause uncontrollable replication. Cancers are a result of genetic mutation. • Investigate and analyze data on prevalence and mortality rates of most common types of cancer. • Ask students to develop pie charts as a result of their analysis.
Evaluate	*Closing questions* • Support or refute the claim: A mutation is always harmful. • What is the probability of Patient #6, who has Huntington's disease, and his wife, who does not, having a child with Huntington's? • Use pie charts to showcase the data of your state and the country for the top three new cancer cases.

9. A lesson plan on speed, distance, and time

Below in Table A.8 we provide a sixth-grade lesson that uses robotics to teach about speed, distance, and time [21]. It tasks students to operate a robot for a given time period on different terrains and record the distance traveled by it. Next, the students compute the speed achieved by the robot for different terrains and get familiarized with the concept of friction. The lesson uses the 5E instructional model and is aligned with the 3D NGSS [1] and CCSS for English Language Arts and Mathematics. It showcases the use of robots in helping students understand the abstract concept of relationship between speed, distance, and time.

TABLE A.8 Lesson plan on speed, distance, and time

Objectives: Using a robot programmed to be driven for a specified time and motor speed setting, students will be able to investigate the resulting linear relationship between distance and time. Next, they will be able to analyze how driving the same robot on surfaces of different material or texture affects the distance traveled by the robot. **Key questions:** What are the important factors affecting energy transformation? **Assessment evidence:** The students illustrate conceptual understanding to modify their robot to complete a given task. They also answer end of lesson exit slip. **Material required:** A pre-programmed LEGO EV3 robot, meter stick, timer, material to make surfaces of different texture for the robot to drive on, and worksheet.

Activity setup: The students are tasked to analyze the relationship between speed, distance, and time by observing, measuring, and recording the distance traveled by a robot that is programmed for a specified time and motor speed setting. By examining the results for the same robot traversing over different terrains, the students will be introduced to the idea of friction.

NGSS Performance expectations [1]:

MS-PS3–1: Students who demonstrate understanding can construct and interpret graphical displays of data to describe the relationships of kinetic energy to the mass of an object and to the speed of an object.

Science & Engineering Practices (SEPs)	Disciplinary Core Ideas (DCIs)	Crosscutting Concepts (CCCs)
Analyzing and Interpreting Data Analyzing data in 6–8 builds on K–5 and progresses to extending quantitative analysis to investigations, distinguishing between correlation and causation, and basic statistical techniques of data and error analysis. Construct and interpret graphical displays of data to identify linear and nonlinear relationships.	**PS3.A: Definitions of Energy** Motion energy is properly called kinetic energy; it is proportional to the mass of the moving object and grows with the square of its speed.	**Scale, Proportion, and Quantity** Proportional relationships (e.g., speed as the ratio of distance traveled to time taken) among different types of quantities provide information about the magnitude of properties and processes.

Common Core

English Language Arts/ Literacy: RST.6–8.7

Mathematics: 6.EE.A.2, 7.EE.B.3, 7.EE.B.4, 8.EE.5, 8.EE.6

	Unit Plan
Engage	• Showcase various images to students and ask them to observe things they notice about energy. For example, skating, flat tire, a roller coaster. • Ask the students to discuss what happens to their bodies when they sit on a roller coaster or how they can prevent a flat tire.
Explore	*Robotics-aided activity* • Situate the robot at the 0 cm mark of a meter stick placed on a table. • Run a robot program designed for a fixed motor speed with five different time periods and record the distance traveled corresponding to each time period in Table A.8.a. • Ask the students to use the collected data to draw a line graph and use the slope formula to determine the speed of robot.

(*Continued*)

TABLE A.8 (Continued)

TABLE A.8.a Students record distance traveled by robot on the surface of table

Time (seconds)	Distance (cm)
0	
2	
4	
6	
8	
10	

- Ask the students to repeat the same activity on a different surface (mulch, bumpy candy path, etc.) and record the resulting data in Table A.8.b.
- The students then graph their new data for each different experimental surface.

TABLE A.8.b Students record distance traveled by robot on an experimental surface

Time (seconds)	Distance (cm)
0	
2	
4	
6	
8	
10	

Explain

Discussion and explanation of new material

- Based on the resulting graphs, discuss what would have caused differences in the line graphs of the two activities on different surfaces.
- Discuss the impact of changing the wheels of the robot or the mass of the robot.
- Explain the terms friction, rate of change, acceleration, and friction.
- Share real-life examples.

Elaborate

Robotics-aided activity

- Brainstorm ways to modify the robot to be successful in traveling a given path on the experimental surface in the shortest amount of time. This may include change in the robot program or change of robot tires.
- Showcase your ideas and run the new experiments.

Evaluate	*End of lesson exit slip* • How much time does it take for a car traveling at 65 mph to cover a distance of 325 miles? • How fast is Joe's throw if he can throw a football to a distance of 150 ft in 2.5 seconds? • Two cities M and N are 340 miles apart. A car starts at 10 AM from M toward N and travels at 80 mph. Another car starts at 1 PM from N and travels toward M at 20 mph. At what time do they meet?

References

1. NGSS, *Next Generation Science Standards (NGSS): For States, By States*. 2013, Washington, DC: The National Academies Press; Available from: www.nextgenscience.org/.
2. Common Core State Standards Initiative, *Common Core Standards for Mathematics*. 2010, Common Core Standards Initiative; Available from: https://learning.ccsso.org/wp-content/uploads/2022/11/Math_Standards1.pdf.
3. Collins, A., Joseph, D., and Bielaczyc, K., Design research: Theoretical and methodological issues. *The Journal of the Learning Sciences*, 2004.13(1): p. 15–42.
4. The Design-Based Research Collective, Design-based research: An emerging paradigm for educational inquiry. *Educational Researcher*, 2003.32(1): p. 5–8.
5. Kelly, A., Design research in education: Yes, but is it methodological? *The Journal of the Learning Sciences*, 2004.13(1): p. 115–128.
6. Moorhead, M., Listman, J.B., and Kapila, V., A robotics-focused instructional framework for design-based research in middle school classrooms, in *Proceedings of ASEE Annual Conference and Exposition*. 2015, Seattle, WA; Available from: https://peer.asee.org/23444.
7. Wang, F. and Hannafin, M.J., Design-based research and technology-enhanced learning environments. *Educational Technology Research and Development*, 2005.53(4): p. 5–23.
8. Barab, S. and Squire, K., Design-based research: Putting a stake in the ground. *The Journal of the Learning Sciences*, 2004.13(1): p. 1–14.
9. Anderson, T. and Shattuck, J., Design-based research: A decade of progress in education research? *Educational Researcher*, 2012.41(1): p. 16–25.
10. Iivari, J. and Venable, J., Action research and design science research—Seemingly similar but decisively dissimilar, in *Proceedings of European Conference on Information Systems*. 2009, Verona, Italy. p. 1642–1653.
11. You, H.S., Chacko, S.M., and Kapila, V., Examining the effectiveness of a professional development program integration of educational robotics into science and mathematics. *Journal of Science Education and Technology*, 2021.30(4): p. 567–581.
12. You, H.S. and Kapila, V., Effectiveness of professional development: Integration of educational robotics into science and math curricula, in *Proceedings of ASEE Annual Conference and Exposition*. 2017, Columbus, OH; Available from: https://peer.asee.org/28207.
13. Schmidt, D.A., Baran, E., Thompson, A.D., Mishra, P., Koehler, M.J., and Shin, T.S., Technological pedagogical content knowledge (TPACK): The development and

validation of an assessment instrument for preservice teachers. *Journal of research on Technology in Education*, 2009.42(2): p. 123–149.

14. Carberry, A.R., Lee, H.-S., and Ohland, M.W., Measuring engineering design self-efficacy. *Journal of Engineering Education*, 2010.99(1): p. 71–79.
15. Rahman, S.M.M., Krishnan, V.J., and Kapila, V., Exploring the dynamic nature of TPACK framework in teaching STEM using robotics in middle school classrooms, in *Proceedings of ASEE Annual Conference and Exposition*. 2017, Columbus, OH; Available from: https://peer.asee.org/28336.
16. Rahman, S.M.M., Chacko, S.M., Rajguru, S.B., and Kapila, V., Fundamental: Determining prerequisites for middle school students to participate in robotics-based STEM lessons: A computational thinking approach, in *Proceedings of ASEE Annual Conference and Exposition*. 2018, Salt Lake City, UT; Available from: https://peer.asee.org/30549.
17. Krishnan, V.J., Rajguru, S.B., and Kapila, V., Analyzing successful teaching practices in middle school science and math classrooms when using robotics (Fundamental), in *Proceedings of ASEE Annual Conference and Exposition*. 2019, Tampa, FL; Available from: https://peer.asee.org/32092.
18. Ghosh, S., Rajguru, S.B., and Kapila, V., Investigating classroom-related factors that influence student perceptions of LEGO robots as educational tools in middle schools (Fundamental), in *Proceedings of ASEE Annual Conference and Exposition*. 2019, Tampa, FL; Available from: https://peer.asee.org/33023.
19. Bybee, R.W., Taylor, J.A., Gardner, A., Scotter, P.V., Powell, J.C., Westbrook, A., and Landes, N., *The BSCS 5E Instructional Model: Origins and Effectiveness*. 12 June 2016, Colorado Springs, CO: Office of Science Education, National Institutes of Health.
20. Ahmed, S., Ghosh, S., Lam, J., and Trivedi, D., *NGSS Lesson Plan: Genetic Mutations*. 2018, NYU-DRK12: Teaching STEM with Robotics; Available from: http://mechatronics.engineering.nyu.edu/pdf/genetic-mutations.pdf.
21. Lam, J., Ahmed, S., and Trivedi, A., *Scaffolding in Instruction*. 2018, NYU-DRK12: Teaching STEM with Robotics; Available from: http://mechatronics.engineering.nyu.edu/pdf/friction.pdf.

APPENDIX B

Online repository of robotics-based lessons

1. LEGO robotics-based lessons

Under an NSF-funded DR K-12 project, our team of doctoral and postdoctoral researchers as well as middle school science and math teachers co-designed and pilot-tested the following LEGO robotics-based science and math lessons aligned with the Common Core State Standards. These lessons can be accessed at the following link: http://mechatronics.engineering.nyu.edu/k12-stem/drk-12/lessons.php#tab_sec1. See also Table B.1.

2. NGSS-plus-5E robotics lessons

Under an NSF-funded DR K-12 project, since 2018, our team of doctoral and postdoctoral researchers as well as middle school science and math teachers co-designed and pilot-tested the following LEGO robotics-based science and math lessons, aligned with the Next Generation Science Standards (NGSS) and framed under the 5E instructional framework. These lessons can be accessed at the following link: http://mechatronics.engineering.nyu.edu/k12-stem/drk-12/lessons.php#tab_sec2. See also Table B.2.

3. Additional robotics-based STEM lessons

Under an NSF-funded GK-12 STEM Fellows project, our team of engineering students collaborated with K-12 STEM teachers to design, prototype, and test 75 STEM lessons that have been published on ***Teachengineering.org***. A complete list of these lessons is available at: http://mechatronics.engineering.nyu.

edu/resources/k12-stem/amps-cbri/classroom-activities/teachengineering.org.php. In Table B.3 we list 30 LEGO robotics-based sample lessons that not only address relevant K-12 science and math curriculum standards, they also allow students to see how the disciplines of science, technology, engineering, and mathematics connect with and draw from each other.

TABLE B.1 Illustrative LEGO robotics-based science and math lessons

1.	Center of Mass Activity for a Unit on Forces and Motion on Earth
2.a.	Rubber Band Robot Build Lesson for a Unit on Energy
2.b.	Collision Activity for a Unit on Energy
3.	Driving Formula Activity for a Unit on Expressions and Equations
4.a.	Distance with Gear Ratio Lesson for a Unit on Ratios and Proportions
4.b.	Gear Ratio and Torque Lesson for a Unit on Ratios and Proportions
4.c.	Gear Ratio and Velocity Lesson for a Unit on Ratios and Proportions
4.d.	Gear Ratio Activity for a Unit on Ratios and Proportions
5.	Data Interpretation of Acceleration Lesson for a Unit on Linear and Nonlinear Relationships
6.a.	Dry Landing Airplane Statistics and Landing Lesson for a Unit on Statistical Data Collection
6.b.	Wet Landing Airplane Statistics and Landing Lesson for a Unit on Statistical Data Collection
7.	Functions of Linear Relationships and Map Skills Unit
8.	Number Line Lesson for a Unit on Integers
9.	Sustaining Life on Mars Lesson for a Unit on Earth Science and Pythagorean Theorem
10.	Tug of War Lesson for a Unit on Physical Science
11.	Cell Cycle Lesson for a Unit on Cell Division
12.	Constant of Proportionality Lesson for Unit on Proportional Relationship
13.	Energy and Simple Machines Unit
14.	Energy Transfer, Surface Area, and Volume Unit
15.	Finding the Unit Rate Lesson for a Unit on Ratio and Proportional Relationships
16.	Newton's 2nd Law – Law of Acceleration Lesson for a Unit on Force and Motion
17.	Least Common Multiple Lesson
18.	PAC-MAN Translation Lesson for a Unit on Translations
19.	Percentage Lesson
20.	Robotic Fish Population Tester Lesson
21.	Temperature Lesson for a Unit on Measurement

TABLE B.2 Illustrative NGSS-plus-5E robotics lessons

1. Scale of Solar System
2. Forces and Interactions
3. Antibiotic Resistance
4. Friction
5. Genetic Mutations
6. Animal Cell
7. Cell Transportation
8. Echolocation
9. Energy
10. Engineering and Energy Transformations

TABLE B.3 Illustrative robotics-based automated lab apparatuses and devices for K-12 STEM activities

Lesson	Link
Elementary School Lessons	
1. Let's Take a Slice of Pi	www.teachengineering.org/activities/view/nyu_pi_activity1
2. On-Track Unit Conversion	www.teachengineering.org/activities/view/nyu_unitconv_activity1
3. Parallel and Intersecting Lines—A Collision Course?	www.teachengineering.org/activities/view/nyu_parallel_activity1
4. Robo Clock	www.teachengineering.org/activities/view/nyu_roboclock_activity1
5. Robot Wheels!	www.teachengineering.org/activities/view/nyu_robotwheels_activity1
6. Robotic Perimeter	www.teachengineering.org/activities/view/nyu_robotic_perimeter_activity1
7. How Cold Can You Go?	www.teachengineering.org/activities/view/nyu_howcold_activity1
8. How Fast Does Water Travel Through Soils	www.teachengineering.org/activities/view/nyu_permeability_activity1
9. Save the Stuffed Animal! Push and Pull	www.teachengineering.org/activities/view/nyu_pushpull_activity1
10. The Claw	www.teachengineering.org/activities/view/nyu_claw_activity1
11. Wide World of Gears	www.teachengineering.org/activities/view/nyu_gears_activity1
Middle School Lessons	
1. About Accuracy and Approximation	www.teachengineering.org/activities/view/nyu_accuracy_activity1
2. Discovering Phi: The Golden Ratio	www.teachengineering.org/activities/view/nyu_phi_activity1
3. The Fibonacci Sequence and Robots	www.teachengineering.org/activities/view/nyu_fibonacci_activity1
4. Accelerometer: Centripetal Acceleration	www.teachengineering.org/activities/view/nyu_accelerometer_activity1
5. Measuring g	www.teachengineering.org/activities/view/nyu_measuring_activity1
6. Measuring Pressure	www.teachengineering.org/activities/view/nyu_measuringp_activity1
7. Runaway Train: Investigating Speed with Photo Gates	www.teachengineering.org/activities/view/nyu_train_activity1
8. Test-A-Beam	www.teachengineering.org/activities/view/nyu_beam_activity1
9. The Balancing Act	www.teachengineering.org/activities/view/nyu_lightbalance_activity1
10. You've Got Triangles!	www.teachengineering.org/activities/view/nyu_triangles_activity1
11. Biomimicry: Echolocation in Robotics	www.teachengineering.org/activities/view/nyu_biomimicry_activity1

High School Lessons

Lesson	Link
1. Means, Modes, and Medians	www.teachengineering.org/activities/view/nyu_mmm_activity1
2. Arctic Animal Robot	www.teachengineering.org/activities/view/nyu_arctic_activity1
3. How to Pull Something Heavy	www.teachengineering.org/activities/view/nyu_heavy_activity1
4. Measuring Distance with Sound Waves	www.teachengineering.org/activities/view/nyu_soundwaves_activity1
5. Molecules: The Movement of Atoms	www.teachengineering.org/activities/view/nyu_molecules_activity1
6. Projectile Motion	www.teachengineering.org/activities/view/nyu_projectile_activity1
7. The Science of Spring Force	www.teachengineering.org/activities/view/nyu_springforce_activity1
8. Tug of War Battle Bots	www.teachengineering.org/activities/view/nyu_activity1_battlebots

INDEX

Note: Page numbers in **bold** indicate a table.

Printed in the United States
by Baker & Taylor Publisher Services